Teach Math Like This, Not Like That

Teach Math Like This, Not Like That

*Four Critical Areas to
Improve Student Learning*

Matthew L. Beyranevand

ROWMAN & LITTLEFIELD
Lanham • Boulder • New York • London

Published by Rowman & Littlefield
A wholly owned subsidiary of The Rowman & Littlefield Publishing Group, Inc.
4501 Forbes Boulevard, Suite 200, Lanham, Maryland 20706
www.rowman.com

Unit A, Whitacre Mews, 26-34 Stannary Street, London SE11 4AB

British Library Cataloguing in Publication Information Available

Library of Congress Cataloging-in-Publication Data

978-1-4758-3360-7 (cloth : alk. paper)
978-1-4758-3361-4 (pbk. : alk. paper)
978-1-4758-3362-1 (electronic)

♾ ™ The paper used in this publication meets the minimum requirements of American
National Standard for Information Sciences Permanence of Paper for Printed Library
Materials, ANSI/NISO Z39.48-1992.

Printed in the United States of America

Contents

Foreword

I am a former middle and high school mathematics teacher and now a professor at Harvard University at the Graduate School of Education. My research is focused on mathematics teaching and learning, especially in the area of algebra. I also provide a lot of professional development for teachers around algebra. A couple of years ago, I received a large grant from the National Science Foundation to evaluate some instructional materials that colleagues and I had developed for improving the teaching of Algebra I. I approached Matthew Beyranevand about initiating a collaboration with middle and high school teachers in Chelmsford, Massachusetts—where Matthew was working as the K–12 mathematics coordinator—as part of my project.

Because of the nature of my research and the professional development that I provide, I spend a lot of time with mathematics teachers and district mathematics coordinators around the United States and world. As I got to know Matthew, it became clear that he is a one of a kind—someone who is passionate about mathematics, about teaching mathematics, and about sharing his excitement about mathematics with students, teachers, parents, and community members. Matthew is not only energetic, dynamic, creative, and entertaining; he is also approachable, thoughtful, and friendly. In addition, he is a respected leader in his district and state.

Anyone who has ever worked with Matthew or watched "Math with Matthew"—his wacky and delightfully entertaining Web series on mathematics for children—will very clearly see that this book reflects who Matthew is as a person and an educator, in several ways.

First, this book is *comprehensive*. It touches on just about everything a mathematics teacher should be thinking about before, during, and after instruction. Clearly Matthew has spent a lot of time thinking deeply about exactly what teachers need to know, as is evidenced by the chapters on planning, assessment, pedagogy, and relationships.

Second, this book is *actionable*. Open the book to any random page and begin reading—it is immediately clear that this book is extremely accessible and at the same time full of easily implementable recommendations. Matthew has compiled a large collection of great ideas for big and small ways that math teachers can improve their teaching.

Third, this book is *comprehensible*. The format of the book—the "Teach like this"/"Not like that" structure—is creative yet also readable, helpful, and tremendously practical, with concrete suggestions and vignettes that are easy to engage with and use. This book offers deep, thought-provoking, yet easily understandable suggestions for improving mathematical practice.

And finally, this book is, like Matthew, *approachable*. The book is written by a teacher and for teachers, in an engaging and conversational format. When reading this book, you feel like Matthew is speaking directly to you. It is full of encouragement, specific suggestions, and even a bit of fun. A teacher with a preparation period can read sections of the book in a short period of time, gleaning suggestions that can be implemented during the next class period.

Matthew is that one mathematics teacher who all of his students will always remember. He has put his heart and soul into this book, and I'm confident that the book will positively impact all who read it.

—Jon R. Star, PhD, professor of education, Harvard University,
Cambridge, Massachusetts

Acknowledgments

This book could not have been written without the love and support of my wife, Valerie. Her insight as an accomplished educator helped me develop my thoughts on each of topics in the book as well as provide me the time and support to write. Thank you to my editor, Sarah Jubar, who consistently afforded me the guidance and feedback necessary to complete this project. I am appreciative to my readers Lombro Ristas, Sheila Solly, Hilary Kreisburg, and Robyn Cutright who were instrumental in the edits and extension of ideas. Thank you to Professor Jon Star for writing the foreword and being a role model for me and all math educators. Finally, I offer my appreciation to Professor Regina Panasuk, who has been my mentor for almost twenty years.

Introduction

Teaching mathematics is one of the most difficult and important jobs that anyone can do. Mathematics is a critical part of education and an essential building block for problem solving skills that are needed in the real world. However, many students struggle to learn and understand mathematical concepts, and as teachers we need to do everything we can to help our students learn. This book focuses on four areas necessary to be an impactful teacher of mathematics: planning, pedagogy, assessment, and relationships.

This is a book for *any* math teacher at *any* level, new teachers and veteran teachers alike. We are all learners no matter how much experience we have. Drawing on research and practical experience with implementation, it examines over thirty ideas from mathematics education and shares what to do and what not to do.

This book is designed to be easy to use, easy to read, and easy to implement. The goal is not to use all the buzzwords in education in a verbose, grandiose way. Rather than making you spend your time interpreting academic prose, this book is intended to help you reflect on your current practice with each topic. If you haven't thought much about a particular area of math education or know you struggle with it, this book will show you simple tweaks, additions, and practices to implement into your classroom.

In order to improve student learning, this book suggests that you take a four-pronged approach that incorporates the best strategies for planning, pedagogy, assessment, and relationships. Mindful planning takes care of a lot of woes in the classroom. You need a solid plan in place to identify the standards, choose the curriculum you will use, develop a reasonable timeline to accomplish each objective, and allow a period of reflection not only for yourself but also for your students.

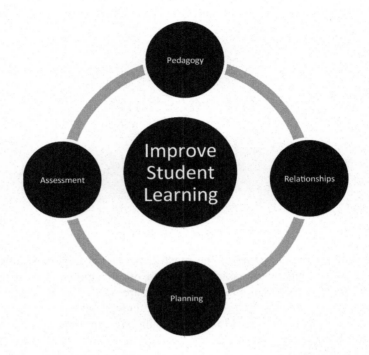

Matthew Beyranevand.

Pedagogy, or the art of helping students learn the concepts, is also a critical component to successful teaching. Understanding and utilizing the best research-based methods will help alleviate your frustration when students "don't get it." However, to quote the famous author Maya Angelou, "Nothing will work unless you do." Your top priority is to be the model learner that you expect your students to be. Your teaching should reflect a continuous cycle of learning, implementing, reflecting, adjusting, and starting over. Education is fluid in that it is continually changing as student needs, college and career needs, standards, and technology change. Although we might not agree with all the things we are being asked to do, having the right mindset will allow us to remain happy and productive in our jobs.

Assessment means so much more than just giving a test for a grade on the report card. This book will help you to develop formative and summative assessment strategies with a focus on formative assessment, since summative assessments are typically included in whatever curriculum you happen to be using. Summative assessment gives us a snapshot of how well students have absorbed and retained information. Formative assessment, however, drives how and what we teach on a daily basis to a diverse group of students. The truly great teachers instinctively use formative assessment to be able to meet the needs of students on a continual basis.

Maya Angelou also said, "I've learned that people will forget what you said, people will forget what you did, but people will never forget how you made them feel." Our relationships with students directly define our success with them. It is not our job to be a "friend" to students, but if they know that you truly care about them and their success, it is amazing what they will do for you on days they don't want to do it for themselves. When you put all these ideas into action you will be well on your way to becoming the teacher that you were destined to be.

For each of the ideas presented in the book, a brief introduction will be shared and then two different perspectives will be detailed with examples. The first is *Not Like That*, which is often the traditional way of teaching mathematics or the less effective approach. This is how many of us approach the concept explained and in many cases, the way I approached it when I began as a middle school teacher almost twenty years ago. The second perspective is *Teach Like This*, which is my recommended approach based upon research and my own experience as a teacher, math coordinator, and graduate instructor of math education. This is what we should be doing to be more effective and efficient and to help improve our students' learning of mathematics.

Chapter One

Planning

The time spent in the classroom helping students really understand mathematics is what truly makes someone a teacher. The time spent planning and preparing for those precious minutes is what makes someone an effective teacher. Planning the phases of a lesson, determining the content standards and objectives, creating appropriate modifications, and collaborating with other math and special education teachers are just some of the critical parts of teaching math. This chapter details what has worked (Teach Like This!) as well as what has not worked (Not Like That!) for each of the different aspects related to planning.

PLANNING LESSONS

Formalized lesson planning is a highly emphasized skill in teacher education programs. Aspiring teachers are required to create and implement many lesson plans in order to acquire a teaching license or certification. However, there are no agreed upon guidelines or formats for creating lesson plans. These plans usually take hours to create and often provide less-than-stellar results when implemented. A simple Internet search will show hundreds of different styles of lesson plans in mathematics and other subjects. However, certain approaches are more effective and less time consuming for teachers, who never have a moment to waste.

Not Like That!

Teachers all have some sort of set curriculum that we use to instruct our students. For some it is a modern curriculum with support systems, for others it is a textbook to take us through the year, and still others develop their own mathematics curriculums. Whatever the curriculum is in your school and classroom, it is still essential to plan your daily lessons.

Mr. Jones is a very traditional teacher who relies heavily on his textbook to make decisions for him. He starts the academic year on page 1 of the book and sees how far he can go before the end of the school year. The topics in the first six chapters always get covered, while the topics in the last few are never introduced. This approach does not represent effective planning, and the idea of using only the material in the textbook to guide your instruction is not always the best practice. It is not reasonable to take the textbook for granted and assume that it covers all the standards required, and no curriculum "knows" your students and their needs, background knowledge, or specific learning styles like you do.

Teach Like This!

Realistic and practical lesson planning differs significantly from the beliefs shared in education course work. Most universities in the United States require students to create extensive lesson plans that take multiple hours to complete as part of their licensure process. There is value in knowing how to create and implement these formalized plans, but as teachers transition to the classroom, very few actually *write out* the same plans on a daily basis. The process often shifts to the other extreme, and teachers simply outline the rough details in a plan book rather than making formalized lesson plans. This is not a negative indictment of teachers, but rather indicative of the significant time commitment required to create the formal plans.

Using the foundation of Grant P. Wiggins and Jay McTighe's (2005) *Understanding by Design* and Regina Panasuk's *Four Stages of Lesson Planning* (2002), lessons should not be driven by an activity, but rather by the desired outcomes that demonstrate the ability to answer essential questions. Math teachers may find the following format to be a quick and effective method to plan lessons. Please note that the order in which the lesson is prepared is *not* the same in which it is executed.

Lesson Purpose: Identify the state standard and then the objective or Essential Question associated with the standard. Students should be able to answer the Essential Question at the conclusion of the lesson; this idea will be developed further in chapter 2. The state standard and objective should be the first two items identified. In other words, by the end of the class period, what should the students know and be able to demonstrate?

Homework: Determine the in-class work or homework assignment that students should be able to complete based upon the material from the class period. If homework is to be assigned, it should be aligned to the objective and standard as well as any prior knowledge necessary to complete the assignment. Homework problems should not be an afterthought, but selected with a clear purpose that practices, reinforces, or extends thinking from the class period.

Learning Experience: Now it is time to determine the activities that must occur to meet the objective, allow the students to answer the Essential Question, and have the knowledge to be able to complete the homework problems. It is also important to identify the necessary vocabulary that you will have to cover. Depending on the length of the class period, there should ideally be multiple phases to the learning experience, including both a teacher-centered portion and a student-centered portion. Formative assessments, which are quick checks for understanding, should be built into the lesson to make sure that learning is occurring.

Warm Up: Through either mental math activities or having students complete problems shared via projector or handout, provide an opportunity to have students practice and recall the prior knowledge necessary for the day's lesson. This also gives you immediate information on whether the class or individual students do not have the necessary content knowledge to be successful in the lesson.

This whole process of designing this lesson plan can be completed in about fifteen minutes and helps teachers plan in an effective manner. Below is an example of what this could look like from a sample sixth-grade lesson:

Table 1.1. Sixth-Grade Math Lesson: Three Dimensional Figures Lesson Purpose

Lesson Purpose	Standard: CCSS.MATH.CONTENT.6.G.A.4 Represent three-dimensional figures using nets made up of rectangles and triangles, and use the nets to find the surface area of these figures. Apply these techniques in the context of solving real-world and mathematical problems.
Essential Questions	What are some of the methods for determining the surface area of a prism? What are the characteristics of a prism?
Homework	P. 328, Questions 7-9, 28. First three problems are drawing two dimensional representations of prisms and the last question is a tent and students asked to find the amount of material needed to make the tent.
Learning Experiences	Share with the students in small groups (3–4 students) examples of rectangular and triangular prisms and have them unfold them to observe the net. Follow Activity 2 from Section 8.2. Circulate, check answers. Next have students use grid paper to draw nets for the prism and observe the characteristics and find the surface area. Follow Activity 4 from 8.2. Have student give thumbs up, sideways, or down on understanding.
Warm Up	Review of vocabulary terms associated with a prism.

Newer teachers of mathematics have frequently expressed frustration with how impractical and time consuming the traditional lesson planning template can be. This simplified version has all the essential components of an effective lesson plan and encourages teachers to plan in a format that has been proven successful through concrete application within schools.

KNOWING THE CONTENT

It is important for mathematics teachers to know not only the grade-level content they teach, but also the content for at least a few years beyond this level. This is usually not a problem for high school mathematics teachers, as most have an undergraduate degree in this field and are fully certified to teach mathematics.

However, only 23 percent of middle school math teachers have an undergraduate degree in mathematics and if a degree in mathematics education is included, the percent rises to just 35 percent (Banilower et al., 2013). Many sixth-grade teachers teach mathematics under a general K–6 certification, which requires almost no formal mathematics training.

Elementary school teachers have the most difficult job as they try to become an expert in five or more content areas, including reading, writing, science, and social studies, in addition to mathematics. Most elementary school teachers are literacy specialists and often do not have an extensive content knowledge of mathematics outside of the elementary grade levels. Only 4 percent of elementary teachers have a degree in mathematics or math education (Banilower et al., 2013) and these are the teachers that build the foundation of mathematics for students.

Teachers are the leaders and experts in their classrooms. It is important that they not only have a strong understanding of the mathematical procedures they must teach, but also a strong *conceptual* understanding of how the procedures were created and why they work. Since most mathematics builds upon itself for future courses, it is essential for teachers to know how the topics covered in their course will be used and applied in future courses.

Not Like That!

A seventh-grade class is being introduced to how to calculate the area of a circle. Mrs. Baldi immediately shares the formula for this as $A = \pi * r^2$. She then asks the students to substitute in different distances for the radius of a circle to calculate the area. After twenty of these, she asks students to find the radius, if given the area. A perplexed student asks why this is the formula for finding the area of a circle. Mrs. Baldi responds, equally perplexed, by saying that she gave the students the formula and now they just need to learn how to solve the problem with it.

Teach Like This!

If you are a teacher of mathematics, in any capacity, and you don't feel comfortable with all the mathematics that you teach, as well as the mathematics that the students will take in the following year or two, you should seek out appropriate workshops or coursework to gain this knowledge. If you are a middle school teacher, you don't necessarily need to take a Calculus II course or a course on differential equations, but a course on high school math topics such as Teaching for High School Math Understanding could be very helpful.

Ms. Li, a middle school math teacher, originally had a degree in elementary education, and her math content skills were somewhat weak. She recognized and acknowledged her lack of content knowledge and was able to do something about it. After taking a variety of content courses on algebra, geometry, and the history and development of math concepts, she blossomed into one of the strongest teachers in the department.

Mr. Martinez was a very successful early elementary teacher of ten years who decided to become certified to teach Algebra I at the high school level. He took a two-week intensive Algebra I certification course during the summer and began teaching it in the fall. His biggest obstacle was the content knowledge itself. He felt that his elementary training had prepared him well for the teaching methods; however, he struggled with being able to explain *why* a strategy worked, how to determine what they should already know from the middle school level, and how the objective would lay the groundwork for next level math courses. In other words, Mr. Martinez was missing the "links" that bind the concepts together into a cohesive whole.

Mr. Martinez sought out an Algebra I mentor who had a similar teaching style and excellent student test scores and evaluation scores. As he prepared his lessons, he would ask several teachers in Algebra I and Algebra II for tips to explain how it all connects. When he was not sure how to solve a difficult problem, he was not afraid to ask for help. In addition, he would search for resources on the Internet, focusing particularly on interactive notebooks, which create a process for helping organize your thinking and the different styles until he felt comfortable and confident with the material.

At the end of the year his words were, "This year has been a beast and I feel like a first-year teacher all over again, but I wouldn't change a thing. I started the year with the attitude that I was going to lay my pride aside and return to being a student so I could become the teacher that I want to be." Building the necessary content knowledge helps you to be more secure and comfortable in what you are teaching so that you can focus on the best methods to help your students learn.

IDENTIFYING STANDARDS

Standards are the content in which students should be proficient by the end of the academic school year. Standards are meant to provide consistency between one classroom and another classroom, and from one school to another school, one town to another town, and even one state to another state. All students in fifth-grade spend considerable time learning how to add and subtract fractions. However, how you instruct, how you assess, and the platform that you use for learning are not dictated by any standard.

Not Like That!

Mr. Marks issues a textbook the first day of school. The second day he begins with lesson 1 in chapter 1. As the lesson begins, he reviews the examples cited in the book, has students complete the practice questions, reviews the answers, assigns homework, and then asks the students to check their odd-numbered answers in the back of the book. He either gives the test that is in the textbook or he uses a test generator to create a test. He continues in this pattern for the remainder of the year. This is the way he was taught math in school and the only way in which he instructs his students.

His textbook is the guiding force throughout the year for determining topics, pacing, assessment, and so forth. His goal is to start from chapter 1 and see how much he can complete before the end of the year. He does not make sure that all topics and standards are covered, nor does he identify an essential or "power" standard to focus on. He makes it through eight chapters by the end of the year and feels good about this as his colleague only made it through seven chapters.

Mr. Hazelwood teaches Algebra II. Every year he has students matriculate into his course from four different Algebra I teachers. He knows within the first week which teacher each student has had without asking. How? Because the department does not teach the standards cohesively and each teacher has chosen the topics (not always a *standard*) that they feel are most important, and they have taught them in the manner in which they are most comfortable. This makes him angry because he now has to fill in learning gaps and try to make sure that he is using a teaching strategy that all the students will know. Not only is it frustrating, it is time consuming and takes crucial hours away from teaching his standards.

Teach Like This!

Standards should be the most important foundation of all your plans, activities, homework, and assessments. This is not a subjective determination, such as whether you agree with the Common Core State Standards, but rather an objective determination of the district- or state-approved standards. These must be the ultimate guiding force for what learning should take place over the academic year.

If you find the standards to be overwhelming, worded in a pretentious manner, and vague, you are not alone. However, there are many ways to unpack the standards and determine what learning needs to take place and how to assess progress. Using resources or Web sites, you can search for a particular standard and examine the types of problems from a variety of sources and in a variety of formats. Here is an example for Standard Cluster 6.EE.B on "Reason about and solve one-variable equations and inequalities."

The Jonas family had a really busy day. After leaving their home, Mrs. Jonas dropped son Cody at his tennis practice. She then drove daughter Kristin to her soccer game and stayed to watch. After the game, mother and daughter picked up Cody on the way home. Once home, Mrs. Jonas saw that they had driven 20 miles that day. How far are the tennis courts from home?

a. Use any method you like to find the distance between the tennis courts and home.

b. Set up and solve an equation in one variable to find the distance between the tennis courts and home.

Note: The family car can only travel along the main streets (gridlines), and all distances between cross streets are the same distance. Assume Mrs. Jonas took the most direct routes to and from her destinations.

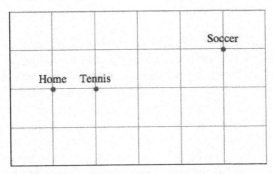

Figure 1.1.   Example problem related to Standard Cluster 6.EE.B. Illustrative Mathematics.

Understanding how the standards are formally assessed is also critical. In a way, this allows you to work "backward" to develop an understanding of the standards. Many states that have standardized testing have also provided information on what standards are major, minor, and supplemental, and some have even given the percentage of questions on the assessment that will address a particular standard. This information could be used to determine

Table 1.2.   Fifth-, Sixth-, and Seventh-Grade Standards Related to Dividing Fractions

Grade level	Standard Related to Dividing Fractions
Fifth Grade	CCSS.MATH.CONTENT.5.NF.B.7.A Interpret division of a unit fraction by a nonzero whole number, and compute such quotients. *For example, create a story context for (1/3) ÷ 4, and use a visual fraction model to show the quotient. Use the relationship between multiplication and division to explain that (1/3) ÷ 4 = 1/12 because (1/12) × 4 = 1/3.*
Sixth Grade	CCSS.MATH.CONTENT.6.NS.A.1Interpret and compute quotients of fractions, and solve word problems involving division of fractions by fractions, for example, by using visual fraction models and equations to represent the problem. *For example, create a story context for (2/3) ÷ (3/4) and use a visual fraction model to show the quotient; use the relationship between multiplication and division to explain that (2/3) ÷ (3/4) = 8/9 because 3/4 of 8/9 is 2/3.*
Seventh Grade	CCSS.MATH.CONTENT.7.NS.A.2.B Understand that integers can be divided, provided that the divisor is not zero, and every quotient of integers (with non-zero divisor) is a rational number. If p and q are integers, then $-(p/q) = (-p)/q = p/(-q)$. Interpret quotients of rational numbers by describing real-world contexts.

the amount of time you devote to each and the order in which you teach the standards.

It also helps to look at the standard for the previous grade level that is vertically aligned with your current standard so that you understand what the students should have already been introduced to. Then look at the standard for the grade level following yours so that you know what the next level is for your students. Understanding this vertical alignment will help you prepare your lessons and assessments in a manner that is the most beneficial to your students and colleagues. Table 1.2 illustrates how standards for the current grade are aligned with standards for the previous and next grades.

The curriculum does not define the standards. Rather, the curriculum is the main tool for addressing the standards. You can utilize supplemental curricula and technology in a way that always comes back to the standard. The beauty of open source materials, as opposed to the textbook, is that you have access to the most current and up-to-date materials designed specifically for a particular standard. Amazon Inspire, for example, is an open education resource platform where teachers can source free learning materials for students from kindergarten to twelfth grade.

As math builds upon itself, it is important to understand the progression of mathematics in order to avoid unnecessary repetition and to ensure that there are no gaps in the academia content. Some of the most successful teams of math teachers meet monthly to review the standards and ensure that they are all being covered, and to agree on the expectations for student knowledge.

TRACING THE HISTORY OF MATH

Most teachers learn about the history of mathematics in an undergraduate course that discusses many famous Western mathematicians. Rarely does any of the history or development of mathematical concepts make its way into the classroom when teachers begin their career. However, mathematics has a rich history that can help build student interest in and understanding of mathematics.

Not Like This!

Mr. Gardner does not value the history of math in his class. He refers to the mathematicians who helped develop important concepts as dead, worth studying only in history classes. When asked about the great feud between Isaac Newton and Gottfried Leibniz, he is disinterested and unaware of the path each mathematician took to develop calculus in the seventeenth century. This would have been a perfect opportunity to grab the students' attention and instill an interest in mathematics.

Teach Like This!

Dr. Panasuk regularly shares a historical perspective on which mathematician or culture developed the concepts students are learning. When teaching a new concept, she explains the methods and approaches historical mathematicians used to solve the problems. For example, the Babylonians were responsible for first developing the "completing the square" method to solve quadratic equations. Their approach is similar to the process taught in modern classrooms, but based on conceptual understanding as opposed to the memorization of the steps.

Dr. Panasuk also shows students how it took many mathematicians a significant amount of time to develop the formulas and equations that they are attributed with creating. Through the process, most mathematicians made mistakes, and so students learn that mistakes are part of learning and creating mathematics. Also, students struggle to find an immediate solution when solving problems, but many mathematicians took many years to solve a single problem. This shows students that math is not about speed, but rather devotion and perseverance.

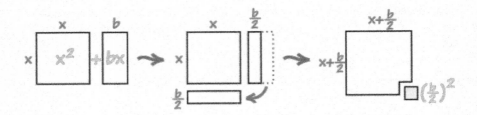

Figure 1.2. The Babylonians method for completing the square. Math Is Fun.[AQ3: provide source for references]

In addition to increasing student interest in learning math, providing a deeper understanding of how the concepts were developed and who was responsible for them also gives teachers added content knowledge. Explaining and showing how the Babylonians and Greeks worked to get more precise values of pi over time only gives Dr. Panasuk additional credibility as a master of her content.

Incorporating the history of math also allows for an obvious interdisciplinary approach to teaching. As we move from a STEM philosophy to a STEAM philosophy (including the arts), the history of math shows us a very obvious relationship between music and mathematics, in particular with Pythagoras, as well as between art and mathematics, such as in geometry. There are many ways to incorporate the history of math into your classroom.

IDENTIFYING MISCONCEPTIONS

A misconception is when a student displays incorrect procedural knowledge while solving a problem. At every level of mathematics there are misconceptions or common mistakes. The more experienced teachers become, the better prepared they will be to anticipate common errors.

Not Like That!

Mrs. Barnes gave a quiz on the distributive property. The whole class did very poorly. She was frustrated with how many students made the same mistake. For instance:

The problem	Students answered
$4(x + 3)$	$4x + 3$

She had shown multiple examples and had given them a homework assignment the night before so they could practice. She started the class by asking if anyone had questions about the homework. The students responded that the assignment was "easy" and turned their homework in to be graded. Therefore, she felt comfortable giving a quiz so that she could move on to the next objective. What went wrong?

Don't assume that students will follow the correct procedures to complete math work just because they have seen or followed correct models. Be prepared for likely misconceptions that students might be having.

It is also important to eliminate ambiguity in language and teaching procedures. Although everyone may be teaching the same standard, the methods and vocabulary used will vary. If you want to have your mind blown, do an Internet search on factoring methods. It is easy to see how students can get easily confused!

Teach Like This!

There are a couple of things that a teacher should be doing to identify misconceptions. First, be on the lookout for common mistakes before you teach the lesson, and be prepared to address those mistakes. Second, give students the opportunities to find and make corrections themselves.

If you are a new teacher or an experienced teacher teaching a new subject that you are not quite comfortable with yet, there are several things that you can do to prepare yourself for the unexpected. Many of the newer secondary math textbooks include notes on the side detailing common misconceptions and are a good source of supplementary material and ideas. The Internet is also a valuable resource as there are many Web sites dedicated to math mistakes at every level.

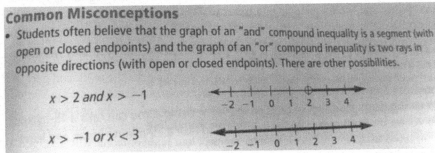

Common Misconceptions

• Students often believe that the graph of an "and" compound inequality is a segment (with open or closed endpoints) and the graph of an "or" compound inequality is two rays in opposite directions (with open or closed endpoints). There are other possibilities.

$x > 2$ and $x > -1$

$x > -1$ or $x < 3$

Figure 1.3.   Example of Common Misconceptions from a textbook. *Big Ideas Math.*

Also, teachers should consider talking openly with the class about misconceptions as a way to deepen students' understanding of the concept. Documenting misconceptions in student math journals and interactive notebooks is a good way to help students remain aware of common misconceptions or mistakes in order to avoid making them in the future.

An effective lesson involving misconceptions comes from an algebra class on factoring polynomials. The teacher has the students identify the misconception and then fix the mistakes that they made. Analyzing other students' work to identify the mistakes and then correctly fixing the mistakes to solve the problem is a much more powerful learning tool than just solving the problem independently.

For new teachers, it is important to collaborate with a mentor or veteran teacher in order to ask about and gain insight into common mistakes made by students for particular concepts. For more experienced teachers, it is always an important practice not just to look at the final answer that students submit, but to analyze their process and thinking to determine fallacies in the way a student is solving problems.

ASSIGNING HOMEWORK

Now for a truly controversial topic: "To give homework or not to give homework." This is one of the newest debates in education. No matter where you are on the spectrum of assigning homework, it is essential to understand some basic guidelines if you decide that it is a necessary part of your curriculum.

Not Like That!

Mrs. Crowson, a veteran elementary math teacher, assigns a worksheet of thirty to forty problems every night for homework. At the beginning of class every day she walks around and gives a 100 or a 0 based on the whether the assignment is complete or not. Students spend at least an hour every night working through the problems. Or do they?

At the middle school and secondary levels students have learned to be quite clever. Many have found the kid who always does the work and gets the answers from him or her. Also, some students form a group and each student does a certain number of problems and then they share the answers. Some have gotten even more creative and have found apps and Web sites where all you have to do is take a picture of the problem and someone will give you a worked-out solution, or one person will air drop it to the entire class. So the question becomes, "whose work are you really grading?"

Mr. Stern, an honors teacher, assigns fewer problems per night but has high expectations for his students and makes the problems very difficult. He includes problems that were often not covered at all in class. These problems require a high level of reasoning skills and the ability to make the jump from solving basic problems in class to applying them in a variety of situations or proving something. He knows that the majority of his students will struggle and most will not be able to complete the assignment. These types of problems are better suited to be done in the classroom under the supervision of the teacher or peers for support rather than isolated on the dining room table.

Teach Like This!

Leaving the debate about *whether* we should assign homework, let's address *why* we assign homework in the first place. Many educators claim that it is what we have always done and what our teachers before us did, so we should do it too. We might feel that we have become a slacker as a teacher if we don't require it of our students. There are times that we need students to spend time outside of school practicing a skill that we no longer have time to work on in class.

So if you do want to assign homework: First, understand the purpose of homework and the importance of selecting appropriate homework that is tied directly to the daily objective or Essential Questions. You should preview the homework assignment during the summary phase of the lesson. Give the students time to understand what the expectations are for their work at home.

Second, homework should just be basic practice of the skills. It should be focused on procedural fluency. The purpose of math homework is to *practice the skills learned in class*. Selecting particular questions to be done for homework is an integral part of daily preparation. The questions selected should have a direct relationship to what occurred during the lesson.

Third, recognize the amount of time that the students will be required to spend on it. For elementary school students, math homework should be no more than ten minutes daily, for the typical student. Twenty minutes is the maximum recommended daily homework in middle school and thirty minutes for high school students. Although we want our students to be the best math students possible, they need to live a balanced life.

Let's look at some possibilities for a homework assignment at the middle and high school levels. If you are a middle school teacher, this is a perfect time for math differentiation. You can give an assignment and have students select the problems they would like to do. For instance, a student must "earn" ten points. There would be five one point problems of one-step equations, five two-step equation problems worth two points each or two problems that have fractions, decimals, or multistep and they are worth five points each. Students must show their work and check their answer using substitution to get credit.

At the high school level, students are starting to take classes based on their ability. For a classroom of students who are academically, behaviorally, or motivationally challenged, assigning homework is an exercise in futility. They simply do not have the resources or desire to do it. Instead, make a contract with them that says they agree to work from bell to bell and there will be no need for homework. The kicker here is that you, as a teacher, have to be prepared to keep them working every minute of that class time. You can head off pleading for "free time" by telling them every day is free; you never charge them for your time.

If you teach a college preparatory class, homework is a game changer. You need to walk the fine line of assigning homework that is challenging, but that most students can successfully complete. Don't assign busy work. Assign real work, which can be completed in a reasonable amount of available time given that these students are often the ones who are actively involved with extracurricular activities. You can assign a few problems that are directly related to aspects of your formally addressed standard in which students are not quite fluent. It is a good practice to give them a final answer to work toward and focus on the process, and not the answer. However, this requires you to grade the *work*, and not the answer.

Ultimately, you have to decide if assigning homework is right for your students. Here are some questions to ask yourself that are both academic and personal in nature:

- What is the purpose of the assignment? Is it necessary? Will it accomplish something that I cannot do during class time?
- Does everyone have access to the resources that will be needed to complete the assignment?
- How long will this take the average student to complete? How long for the students who struggle in class the most? Is this a reasonable amount of time for a child to spend on one activity when they have other classes, extracurricular activities, responsibilities at home, and possibly jobs?
- When and where am I going to grade these assignments? Generally, if you are utilizing class time to grade papers, then you are not teaching or facilitating learning in your classroom. You are shortchanging your students if you are unwilling to take work home.
- Am I going to grade for accuracy and give feedback or is this a completion grade? This can launch into a whole different topic, the validity of grades, which isn't the point here. However, research shows that providing feedback is one of the top-ten most effective teaching strategies. Without feedback, homework is busywork and grading for completion is a responsibility grade and not a grade based on the standards.

You also have to take into consideration the culture of your class, and students' other obligations and responsibilities. Many scholarships now require community service and leadership roles for applicants. This means in addition to academics they have to work on becoming a well-rounded individual that colleges find desirable.

Most teachers believe that homework is a valid learning experience. Take the time to prepare and assess homework with focus and efficiency for both you and your students.

FINDING THE JOY IN MATH

Why are you a current or aspiring teacher of mathematics? One can only assume that you have some love or passion for mathematics. Teaching, in general, but especially of mathematics requires much intrinsic motivation as there are many different careers that can be financially rewarding. However, in observing hundreds of math classes of all levels, it is infrequent to see teachers sharing the joy and beauty of mathematics. Mathematics helps explain the world that we live in and is prevalent in art, music, science, architecture, as well as numerous professions.

Not Like That!

Mr. Buck begins class as soon as the bell rings. Students open their books and class begins. There are no group activities, interactive learning games, or mathematical discourse on who has the correct answer. He does not pose any intellectually stimulating questions to create curiosity. Mr. Buck is a very competent mathematician, but lacks a sense of humor. He knows that students must pass his class in order to graduate, and therefore he does not want to "waste time" sharing the beauty and joy of mathematics through exploration and activities.

Every student is required to take math whether they want to or not. They may not be good at it, enjoy it, or see a need for it in their immediate future, yet we know that the underlying skills that are developed are used inherently in our everyday lives.

You should avoid conveying the attitude that it doesn't matter whether they like it or not and that since they have to take it, they do not have to work at making it enjoyable. Students typically love their elective classes (and teachers) because they *choose* them and have a strong interest in those subjects.

Arts, technical, and humanities teachers may even have to vie for students to enroll in their classes in order to have the numbers required to keep their class funded. Math teachers, on the other hand, know that no matter what, students must take eleven or twelve years of math to graduate, and because of that, in many math classes there is little joyful learning.

Teach Like This!

Mathematics is the study of quantitative patterns in our world. Mathematics can help us explain and understand so much of the world in which we live using numbers, patterns, and relationships. Although it is important to have standards-based learning within your daily lessons and your yearlong plan, it is critical that student learning experiences occur in a joyful and exciting manner.

For instance, consider what occurs within science instruction. Science classes are full of experiments, hands on experiences, and posing questions. Not everything has to be tied into the real world, but you can get students engaged by showing information and posing a question or asking them to pose a question. Similar to a science class, try posing a question about an occurrence, a situation, or some presented data and have the students try to answer the question through collaboration, discussion, learning, and applying mathematics related to the lesson. Students need to view mathematics as an active and engaging experience, not as passive drudgery.

Although the objective may be a little dry, you can also spice it up by being creative in how you present the material and the activities that you use for practice. The Internet is full of delightful activities that allow students to enjoy the *process*, if not the material. For instance:

- Scavenger hunts where students work a problem then search the room for an object covering the answer, then work the problem underneath it until all the problems have been worked correctly.
- Koosh ball activities where the students throw a ball and hit a "circle" and then work the hidden problem. The first person who gets it right gets to throw the ball.
- Any activity that encourages a little competition or provides rewards for answering a challenging question, asking a thought-provoking question, or sharing a different strategy with the class.
- Checking answers using QR codes.
- You can keep an old phone or tablet that will work on wifi for those students who do not have access to technology.

If you are in a school that is fortunate enough to have one-to-one technology devices, endless options are at your fingertips and you should utilize that technology to create excitement and enhance student learning. However, if you don't have much technology there are many, many engaging activities that can be done in a way to make math fun. The possibilities are endless and if you lack a little bit in the creativity department, Web sites such as Teacher Pay Teachers and Pinterest have hundreds of fun activities that are designed for students to have a joyful, engaging, and fun-filled math class. The Global

Math Project is working to bring joyful mathematics to all classrooms around the world.

The more excited you are about what you are teaching, the more students will absorb that energy. It also works in reverse; when students come in pumped and excited, the teacher gets as joyful as they are. Joy doesn't always require an activity; it can simply be a pervasive attitude that begins with you, the leader of the class.

GROUPING WITH A PURPOSE

There is much research on the benefits of grouping students and the variety of ways in which it can be done. In fact, there is so much that it can be quite overwhelming. In professional development sessions, it can seem that administration and curriculum leaders want students to work in groups all the time. In reality, "the start small" philosophy is much more manageable and efficient.

Not Like That!

Ms. Wallace was preparing for her evaluation with her principal. She was trying to meet the objectives in her evaluation rubric and grouping was a big category. She found a great group activity online that went along with her lesson plan and just knew that it was going to impress the principal and result in a high evaluation score. As the big day arrived she had picked four-person groups based on the low/medium/medium-high grouping plan. She had written out role cards so that each student would know what they were supposed to be doing. All materials were ready to go and everything was planned to perfection. She had done everything that she was told to do in her professional development about grouping students. Yet the day turned out to be a *disaster.*

The transition from whole class instruction to the group activity was pure chaos. The students had never moved their desks into a four-person grouping structure. Five minutes of class were wasted to get them in position. Students began reading their "role" cards and the questions started. What does this mean? Why am I the note taker? What am I supposed to do?

When she finally got everyone started, she walked from group to group trying to keep students from talking about everything except the lesson. Emily, the class math genius, didn't have the patience to tutor her group. Madison got frustrated with her group and simply did all the jobs herself. Ms. Wallace found herself tied down by a group that had mastered the art of being enabled. She was frazzled and exhausted and didn't make a good transition back into whole class instruction after the group activity.

Do not have your activity phase of the lesson involve group work unless there is a specific reason. Also, as mentioned in the lesson planning discussion, the plan for the activity phase of the lesson and how the students are to be organized should be determined *after* you develop the objective, homework, and assessments. Don't have groups for every activity, or think "oh, I have a group activity for the day." Group activities have the potential to be extremely time consuming for very little return without proper planning and student training.

Teach Like This!

The first step is to arrange your classroom in a manner that is conducive to all types of grouping. You want to be able to walk easily among all groups, which is to your advantage for managing discipline. Rows and columns are boring and they limit your ability to interact with and engage in proximity control with students. Some type of "pod" works well. There are a few ways you can do this:

- Replace desks with tables for students.
- Arrange desks in groups of three or four where student desks are connected.
- Create quad pods where the desks are all facing forward, but there is a noticeable grouping that allows you to walk into between, in front of and behind each group.

The second step is to start short, sweet, and small. During the guided practice portion of your lesson, have students check an answer with a peer partner in their group. Then check with the entire group. If someone has missed it allow one of the peer partners to explain where the mistake was made. Students love this because it allows them to bolster their confidence before being called upon in class by the teacher. This is also a good time for students who don't know each other to get acquainted in a directed manner.

Step three would be to create a formal group with a more extensive lesson that typically requires some type of product: a poster, a PowerPoint, Prezi, a booklet, or a presentation of some kind. You should be prepared to spend time teaching students about the roles and what they mean in the context of the project. You will need a rubric that explains in detail what is expected from the group project and students should understand how it is an extension of the current objective or standard. If it is not designed to allow discovery and enrichment or solidify the objective it is a waste of valuable time.

There are three main ways to determine who is going to be in each group:

- The teacher chooses based on student academic ability, personality, or student interest.
- The student chooses a partner or group based on friendships (or some wise cookies will look for a smarter student to help them).
- Randomly group the students using sticks, cards, apps, and so forth.

You need to determine your standards and objectives for the lesson as well as your formative assessments and homework, and then figure out how the learning will take place during the class period. During the student-centered portion of the lesson, decisions need to be made as to whether the students should be working independently, in pairs, or small groups. Further, if they are working in pairs or small groups, how are those groups formed? Are the students paired up with those of a similar skill level, or is one student stronger mathematically than the other so that you are relying on peer tutoring?

There is no single correct approach, however, it is suggested that you determine whether there will be group work and how students will be grouped based upon the goals and objectives for the lesson. For example, for a lesson about reviewing prior knowledge, it might make sense to have a struggling student paired up with a stronger student to encourage the peer tutoring when necessary.

However, at times it might make more sense to create homogeneous groups that give the more advanced students a chance to extend their thinking and force students that are within the average range to be active participants rather than relying on the stronger students in the group to do all the work. This is also a good time to have a teacher-led group for struggling students. After guided practice, if the majority of the students are prepared to work independently but you still have less than five who are just not getting it, you can pull them together in a group and continue the guided practice until they are ready to work independently.

A note of caution, as theoretically it sounds perfect to pair a very smart student with a struggling student. Gifted students are not always the best choice for peer tutoring. Not only do they tend to be impatient, but their thinking and approach are often different and they make connections and utilize strategies that other students often cannot understand. High-achieving students tend to do the work for the weaker student. This is why it is extremely important to know your students. At the beginning of the year when allowing students to check answers with a partner or a group note which students tend to enjoy the role of "tutor" and which students others tend to be drawn to when needing an explanation, not just an answer.

Students do not instinctively know how to work in a highly effective group. It is not enough to tell them to just get in the groups to solve problems. The teacher needs to be specific as to the roles and expectations of the group. Is one person the note taker, is the next person going to be sharing out the conclusions, is there a timekeeper, should everybody be contributing equally and how you as a teacher are going to monitor the activity?

COLLABORATING

It is not reasonable or practical to ask every teacher to do all of the work in isolation. In most schools, teachers have teams of colleagues who teach the same grade level or course. Working jointly to prepare lessons, share assessments, and test out new ideas is critical to success.

Not Like That!

In the past, every teacher could close their doors and have complete autonomy in the classroom. Teachers could decide what, how, and when they taught any particular objective and how they wanted to assess the learning. They also had three years to get their act together before they were up for professional status. That was typically a wonderful situation for a teacher—or was it?

New teachers are often thrown into a classroom where it feels like they must either sink or swim. Because you are expected to be an expert the minute you enter the classroom it feels as though asking for help somehow makes you less of a "teacher." Not only do you have to prepare materials, you have to actually manage the behavior of a group of students, and no amount of college can prepare you for this!

Teachers also tend to reinvent the wheel over and over because they don't want to (1) ask for help, (2) appear to be a lazy slacker, or (3) appear to be lacking in content knowledge. However, which is a more effective use of your time—spending your time creating materials that someone has surely already created or spending your time collaborating to determine strategies, best practices, or how an objective leads to the next level?

Even worse are those teachers who have been teaching a long time and subscribe to the philosophy that "this is the way I have always done things." They have filing cabinets full of old hand-written tests that they created fifteen years ago and are still using without revision or reflection.

Fortunately, those days are pretty much over. Why? Because an autonomous classroom is not what is best for a teacher, the students, the school, and the district.

Teach Like This!

Do not feel or act as if you are in this alone, or have to do it all by yourself. Nor should you believe that teaching should be done in isolation. Planning and delivering lessons should be a collaborative approach with many different educator stakeholders.

It is imperative that you realize that you are one piece of a very large puzzle. The bigger picture has to be considered particularly in regard to the students. Our job is to be one step in their journey of learning and in order to do that effectively we have to understand their needs and their goals.

Collaborating with other math teachers who teach the same content and sharing ideas, bouncing thoughts and strategies off each other, and planning with each other is an essential part of being a professional. This can make the difference between a good department and a great department. Regular collaboration can allow us to learn different teaching strategies, find common misconceptions among students, and provide us with the best possible resources, whether it is a book, worksheet, assessment, or an online resource.

When you have an entire math department mining for quality materials and strategies and then collaborating so that a seamless vertical learning process occurs, you will increase learning and achievement scores of students and the effectiveness level of all the teachers involved. The impact will be seen throughout your school and district.

Effective co-teaching can only be successful if there is a trusted relationship among the teachers; time allocated for lesson planning, and predetermined roles for executing the lesson. Therefore, it is very important for special education teachers and support staff to play a role when appropriate in both the planning of lessons as well as the delivery.

As the math teacher in the classroom, you take the lead on the content and the different ways to present a concept but a special education teacher can give insight into how a student's diagnosis may affect a child and the modifications that may be needed for successful learning. The sad fact is that many math teachers find themselves in the position of having to teach special needs children without the benefit of training. Therefore, it is extremely important that you utilize your special education teacher to fill in gaps and increase your knowledge of your students.

In the infamous words of Harry Wong, you should "Beg, Borrow, or Steal," especially if you are a new teacher, a veteran teacher who is changing grade levels, a veteran secondary teacher who is changing subjects, or simply a teacher who is overwhelmed by the process of curriculum development.

KEY TAKEAWAY POINTS

- Traditional lesson planning can be very time consuming, but not planning at all will negatively impact your students' learning of the concepts. Plan in an effective and efficient manner, as outlined in this chapter.

- Standards should be the foundation of your instruction. Your foundation should not be the textbook, state standardized tests, or the concept that you enjoy instructing the most. Focus on your national, state, or local standards and try to have your students reach proficiency or better on each of them.
- Identifying and addressing possible misconceptions with students will help minimize the mistakes and frustration. In addition, it will help deepen the students' understanding when they are able to comprehend the misconception and why it will not work.
- Homework for the sake of homework is not beneficial and will likely make the students frustrated. Short of previously learned skills practice that is directly tied to a lesson is an acceptable way to approach homework.
- Math is beautiful, joyful, and magical. Find ways to show your students the love that you have for mathematics. Create activities for students to explore and discover its beauty.
- When creating groups for students to explore or work with mathematics, take the time to determine the best format. Most students have never learned how to effectively work in a group, so it is important to teach this at the beginning of the year.

Chapter Two

Pedagogy

Pedagogy is the art of teaching. In the past, pedagogy in the classroom meant simply teaching the way that we had been taught. It was natural to think that if it worked for us, it would work for our students as well. Unfortunately, this view did not take into account the fact that we work with different types of students and learning styles, which institutions and students may have varied cultural expectations, and that current technology creates new opportunities and challenges.

Teachers' instructional strategies are driven by factors other than just pure pedagogy. Instructional strategies must represent teachers' own personal and philosophical beliefs along with the school's culture and policies. A huge component of sound pedagogy is understanding and responding to students' culture, environment, background knowledge, and experience. Pedagogy should foster academic engagement, support a sense of community, and promote the well-being of the student while remaining consistent with the vision of the school and district.

Effective teachers are successful at using an array of teaching strategies and grouping methods that help students of all levels to achieve growth and understanding. At the heart of pedagogy is the ability to improvise and intercede in a timely manner with a wide variety of students and abilities as the learning is occurring. This chapter develops pedagogical concepts to enhance any math teacher's strategies for student learning.

DEFINING OBJECTIVES AND ESSENTIAL QUESTIONS

Both objectives and essential questions are integral parts of each lesson plan, but they are not the same thing. Objectives are short-term, day-to-day, measurable goals. They are typically a component of the standards and are concrete statements of what the student should be able to do. Essential questions are not aimed at content mastery, but instead encourage students to think about how the objective will apply in a more global manner.

Not Like That!

Have you ever gone on a surprise trip with someone who won't tell you where you are headed? You look for clues, make random guesses, and get increasingly frustrated when answers are not forthcoming. Particularly if you are being led on the "scenic" route and if you are as detail-oriented as most teachers are, you can't enjoy the trip because you are too full of questions: "How long is this going to take? Do I have everything I need? Am I dressed appropriately? Am I going to enjoy the final destination? Why are we going this direction? Is there a faster way to get there? Why are we doing this?"

This is how a student in a classroom can feel when a teacher starts the lesson without giving a clear picture of where the learning is going to lead. Our destination should not be a surprise for the students. Outlining clear objectives and posing essential questions lays a road map for the student that allows them to focus on precise goals. Before you start an activity, you should give students some idea of the purpose of the activity.

Teach Like This!

Let's start by debunking some of the myths about objectives and essential questions. Objectives should be clear, precise goals for what the student will learn, written using action verbs. You can write objectives on the board so that they are visible and clearly understandable to all students (and any administrator who happens to walk through your classroom). So, let's start by looking at the difference between good and vague objectives (shown below) before comparing an objective with an essential question.

Vague Objective	Specific Objective
Understand fractions	Add and subtract fractions by finding a common denominator.
Apply the Pythagorean Theorem	Use the Pythagorean Theorem to find an unknown side length.

Essential questions are a very different prospect and are much more difficult to formulate. These are *not* simply learning objectives written as questions and put on the board every day. They are thought-provoking, intellectually engaging questions that should encourage students to be creative, curious, and think critically. Answering essential questions should require some type of analysis, prediction, or inference. Essential questions differ most clearly from objectives in the fact that they do not have a specific or single correct answer or solution. The table below compares an objective with an essential question.

Objective	Essential Question
Use multiple ways to represent a number including pictorial, symbolic, and objects.	How do I determine the best way to represent a number?
Approximate the area and circumference of a circle.	When is it appropriate to use an approximation? How important are estimations and approximations in real life? Can you share a good example?
Find the unknown side length of a triangle using the Pythagorean Theorem.	What are some real situations in which someone would need to know how to use the Pythagorean Theorem?
Find the measures of central tendency (mean, median, and mode) of a set of data.	How do the media and large companies use data to influence your choices?
Solve the following equation justifying your steps with the properties of equality.	How do I know where to begin when solving a problem?

The standards should be the driving force behind your yearlong planning or scope and sequence. Objectives and essential questions follow from the standards, and become the foundations for individual lessons. Each lesson should have one or two objectives that drive your strategies and activities for that standard. When you are summarizing the objectives, you can also engage students in exploring the essential question.

Curriculum resources often provide you with sample objectives and essential questions, but you should modify those objectives to meet the needs of your students in your class. For example, the curriculum might assume that the students have some prior knowledge on a topic that your students do not have. It would not be realistic to expect students to add fractions with unlike denominators, for instance, if they do not yet understand how to find a common multiple.

Even though research-based practices like using essential questions seem like new innovations, many have been around a long time without any special names. Essential questions are what teachers with good questioning skills have always naturally embedded into their lessons.

SETTING AN AGENDA

By definition, an agenda is a list of items to be discussed at a meeting. For a classroom teacher, an agenda is a detailed plan that represents one of the most critical components of planning for daily lessons, unit lessons, and year-long sequences. Without an agenda teachers and students lose focus, waste valuable time, and may miss critical standards needed for testing.

Not Like That!

Mrs. Brown likes to stand in the hallway and chitchat with other teachers between classes. She frequently enters her classroom late to find students milling around and texting on cell phones. As she comes through the door she has to herd students to their seats and wait for them to get settled. She finally manages to take roll and then has to find her teaching materials.

She begins her review of the previous day's lesson and gets side-tracked by a discussion of the Friday night football game, and then jumps right into the heart of a lesson for which she has planned an excellent interactive activity for the students. Giving notes and working through the guided practice takes much longer than she had anticipated. When she is finally ready to do her activity, students have to get up and gather supplies such as colored pencils and figure out groups. In the middle of the activity the bell rings and everyone scrambles to leave the classroom. Now she will have to start the activity all over again the next day.

At the end of the quarter, Mrs. Brown laments that she never has time to cover all her material and that she is struggling with classroom management.

Have you ever planned to give a quiz at the end of class and students are in the middle of the quiz when the bell rings? You can't just hand out the quiz at the beginning of the class the next day. The quiz at the end of class was designed to be taken while the material was fresh in their minds.

Maybe you know a teacher like this, who has a rough idea of what will occur in class that day but does not take the time to plan each phase of the lesson?

Teach Like This!

Creating and sharing an agenda is important not only for the teacher, but also for the students. As the teacher, you should know what will occur in each phase of the lesson and how long it will take. In the high stakes world of testing in education, it is critical that you remain focused and use class time wisely. You can create fun activities that allow students to move around and break the monotony of just sitting in a desk and working, yet these activities should be an integral part of the lesson so that you don't waste valuable classroom time.

With an agenda you can make modifications as you go to keep within your time limits. If an activity is taking longer than you anticipated, you can use formative assessment techniques to determine if students are proficient enough to do fewer problems in guided or independent practice; for example, you might decide to only assign the odd or even problems rather than the whole set.

At professional development days or faculty meetings, we expect an agenda that will let us know where we should be, what we should be doing, and when. Our students need this too. An agenda helps give students structure and expectations for what they should be doing and when in the lesson things will happen. Teachers should take thirty seconds at the beginning of the lesson to let the students know the plan for the day. This particularly helps students who suffer from anxiety, have ADHD, or fall somewhere on the autism spectrum. For these students, structure is very important. Letting students know the agenda will help keep you, and them, grounded and focused and can forestall classroom management issues.

Having clear expectations for what is going to occur in class can also help avoid behavior management issues. Elementary teachers with inclusive classrooms can really attest to the need for the strong structure that is provided by an agenda. Most young children feel safer and less anxious when they know what is expected of them and when. An agenda can take away the fear of the unknown.

Being prepared in advance and then sharing your agenda with students is a win–win situation. One of our jobs as teachers is to prepare students for college and careers in ways that are not strictly academic. By using an agenda you teach them the value of managing their time and remaining focused and show them a strategy for organizing their lives successfully.

WARMING UP

Is a warm-up really necessary? The best comparison for an academic warm-up is an athletic warm-up. Could you imagine Usain Bolt, an Olympic sprinter who is regarded as the fastest living human, running without a warm-up?

A serious athlete would never show up late to a group workout and just start running, cycling, or swimming without some type of warm-up. Why? Because it is important to be prepared not only physically, but also mentally for the upcoming efforts. Part of the warm-up process includes getting their heads in the game and getting psyched up.

If warming up is how athletes prepare themselves, it makes sense for teachers to prepare students physically *and* mentally for their classroom challenges. Teachers are trying to enhance their students' academic performance just as athletes work to enhance their physical performance.

Not Like That!

Mrs. Harrison introduces her sixth-grade class to the distributive property on Wednesday. When students come in on Thursday she immediately begins the lesson, asking students to solve equations that require the use of the distributive property. Students are frustrated and unable to solve the problems because they are still struggling to use the distributive property. She ends up stopping the lesson, searching for supplemental material, reviewing the distributive property, and then reteaching her lesson on equations.

Teach Like This!

At the beginning of the lesson, teachers should consider sharing a small number of questions for the students to immediately work on. Some teachers call this a warm-up or a bell ringer. This kind of warm-up serves two purposes: it enhances the overall organization of the class, and more importantly it gives students the opportunity to refresh or activate prior knowledge.

Starting in an organized way helps set the tone for the entire class period. Instead of letting the students coming in talking, socializing, fooling around, and consistently needing to spend a minute to get the class organized, you can have a set of warm-up problems on the board or screen. This helps establish a routine and starts the momentum for the class session. This phase of the lesson should take no more than three to five minutes, and can offer the teacher the opportunity to take attendance, check homework, or even just catch her breath.

Warm-ups allow us to activate prior knowledge or actively engage students in the thinking process. Warm-ups can be classified into three major categories: thought-provoking brain teasers, problems that recall prior knowledge, or objective-specific activities.

Brain teasers do not have to relate to the current objective or prior material but they can be riddles, puzzles, or logic problems used to stimulate the brain and encourage thinking outside the box. These kinds of problems help you encourage higher order thinking within your lesson.

If you have taught students how to solve quadratic equations, then you know the pain of going through the numerous methods and seeing students do well on each strategy as long as they only do one method at a time. Then things fall apart when reviewing and testing come around and students have to select their own strategy. This is when you can use a brain-teaser warm-up problem. Tell the students not to work the problems, but to determine the best strategy and have a mathematical reason as to why. This will create some pretty awesome dialogue with which to open your review lesson.

The second category of problems, which recalls *prior knowledge*, is more general. These could be similar to questions from the previous lesson or homework, and will push students to activate what they learned in the last class and help prepare them to extend their thinking for this class period. Warm-up problems could also be drawn from topics previously studied as a way to review and keep fresh important mathematical concepts. This type of spiraling is typically very effective when preparing for exams or end-of-course testing.

The last category, *objective-specific activities*, targets prerequisite skills that students will be using in the topic you are about to introduce. For example, an elementary teacher might have students count by fives, tens, and twenty-fives before introducing a lesson on counting money. A middle school teacher might review slope before graphing lines from an equation. A secondary teacher might review absolute value and inequalities before introducing compound inequalities. This method provides a rehearsal of the upcoming activity.

The intensity and duration of a warm-up should directly reflect the average classroom ability level. Warm-up problems from any of these three categories should not be challenging problems that require significant assistance for many students because: (1) you are using this warm-up time to get organized and conduct administrative tasks; and (2) this way students begin the class period with success.

When designing your warm-up or bell ringer it is important to think about what students need to know to succeed in the lesson. You want to tease out skills required to accomplish the learning for that objective. The warm-up should not necessarily be done for a grade; it should help you understand your students' ability and content knowledge and serve as a formative assessment to determine if students are prepared to move forward or if reteaching or review are required first.

EXPLORING

Teachers link their objective to their interim assessment or homework with the explore phase. This phase of the lesson has many names: activity, development, guided practice, "we do," and explore. As chapter 1 explained, you should plan this phase *after* you have created the objectives and laid out the assessment. This phase links those two elements together.

Not Like That!

After reviewing the previous day's assignment, Mr. Fish shares example problems and the students write down exactly what he writes. He then has the students work out a couple of problems and check answers with a partner. Finally, he assigns a worksheet for independent practice. This same practice occurs each day until the formal assessment on Friday.

Teach Like This!

Following the same procedure every day can get very boring for students and allows teachers to become complacent about student learning. Just because you taught something during a lesson doesn't necessarily mean that the students learned it. Offering a variety of explore activities for each lesson and different options on a daily basis keeps things fresh, not only for the students, but for you as a teacher.

The explore phase is where learning takes place through a variety of methods. Some different parts of the explore phase of a lesson can include lecture, discussion, graphic organizers, advanced organizers, Cornell notes, think-pair-share, peer partners, or small group activities that will focus on collecting information, making predictions, using reasoning skills, testing answers, and developing or strengthening processes for problem solving.

One of the biggest mistakes teachers can make during the explore phase is taking a passive role and thinking that students have to work *without* teacher assistance in order to reap the full benefits of the exercise. However, students often make mistakes that they are unaware of or they get stuck in a particular phase and quit working. You do not want to give them the answers, but you do want to circulate and assist students by giving guidance or prompting when a student is stuck. Some effective questions that teachers can use in the guiding process include the following:

- Why did you make this step?

  Equations: *"Why did you choose to subtract 3x instead of subtracting 5?"*

- What is the first step of the process?

  Equations: *"Let's go back to thinking about the process and not the problem. What is the first step to solving any equation?"*

- What if?

  Operations with fractions: *"What if the denominators are different?"*

- What is the difference between __ and __?

  Operations with fractions: *"What is the difference between adding and multiplying fractions when you're looking at the denominator?"*

- What information do you need?

  Area formulas: *"What information do you need in order to use the formula for the area of a triangle?"*

- How can you tell by looking at the problem which strategy you are going to use?

  Factoring: *"How do you know when to use GCF or factor by grouping?"*

- Why did you choose that particular strategy?

Quadratic equations: *"Why did you choose the quadratic formula as opposed to factoring?"*

- Can you summarize your strategy for me?

  Coordinate plane: *"Summarize for me how we can find the slope of any line in the coordinate plane."*

- What is the purpose of __?

  Solving equations: *"What is the purpose of using the distributive property?"*

- Is that a reasonable answer?

  Quadratic word problems: *"Is it reasonable that it takes -3 seconds for a ball to hit the ground after it was dropped?"*

You should ask yourself a few questions when you're reflecting on the explore phase:

- Am I using the best resource for this activity?
- Am I grouping students advantageously so that everyone benefits from the group effort?
- Am I asking questions that will lead to higher order thinking skills?
- Am I focusing on a final result or am I developing strategic thinking?
- Am I providing students with opportunities to utilize their reasoning skills?
- Am I providing students with opportunities to develop their own strategy?
- Am I differentiating instruction to meet the needs of all learners?
- Am I creating a dialogue with students that encourages verbalizing of their thought processes?

Students become actively engaged in the learning process when they are given the right tools and atmosphere during the explore phase. They begin to develop an understanding of the process itself and reasoning skills regarding different strategies when you encourage them to slow down and think about what they are doing and why.

SUMMARIZING THE LESSON

To summarize, students must be able to extract information, select the important concepts, condense them in a concise form, and then carry them forward as tools for understanding or problem solving. It is one of the most important phases of a lesson, but often overlooked or skipped due to time constraints.

Not Like That!

Mr. Barrett's class is working in their groups on problems while he circulates around and immerses himself in conversations with groups. Suddenly the bell rings and all the students get up to leave. Mr. Barrett bellows out, "For homework tonight, make sure you do problems 1–13 odd from this section," as the students leave the room.

Students have left amid chaos and without a solidifying summary of the day's objective. Some of the students have understood well and are prepared to complete an independent assignment, some can muddle through most of it, and then you have those few who still don't have a clue.

Teach Like This!

Regardless of how the flow of the explore phase goes, it is critical to leave five to seven minutes at the end of the class period for the summary phase. Even (and especially!) if only half the lesson has been completed, the class needs to reunite for a few important topics of discussion.

First, you can use the summary phase to review the objective or essential question for the class period and determine what progress was made through a variety of formative assessments. (These will be discussed in greater detail in chapter 3.) This is also the perfect time to discuss common errors and misconceptions. After students have worked the problems it is also much easier to ask them higher-order questions: How does this relate to what we have learned previously? Who might use this in the real world? Did anyone use a different strategy and still get the right answer? How is this different from what we did yesterday?

The summary phase is also a good time to discuss and preview the homework assignment for the next class period. Have your students open the book to actually look at the homework questions or use the projector to display them and help the students understand what you expect and how the homework directly relates to the learning that has just taken place. This will significantly increase the likelihood that the students will attempt—and successfully complete—the expected homework assignment.

You can give them tips like "Number 7 is a little tricky. Be sure that you read it carefully," or "Remember we talked about a common mistake and this is where that may come into play." Ask them to read the directions and then verbalize them in their own words, especially if they are being asked to do multiple steps or give more than one answer. For instance, in geometry, a student may be asked to write an equation and find the value of x, but then have to use that information to find the measure of an angle. This is also a good time to review any formulas or vocabulary they may need, particularly if your assignment is spiraled and includes previous material.

You can also share how the topic will be extended. For example, an elementary teacher could inform students that finding the least common multiple is the first step in adding and subtracting fractions. A secondary teacher could share how the Pythagorean Theorem is embedded into problems on the SAT or ACT.

Finally, you can take a short thirty seconds to let the class know what will happen the next class period. Students should know if the topic will be continued the next day, which might relieve stress and anxiety for those who are struggling, or if there will be some type of summative assessment such as a quiz or test for which they will need to prepare. Students should know several days in advance (and be reminded daily during the lesson summary) about any type of formal test that is approaching.

In broad strokes your lesson can be broken into three parts: activation, problem-solving, and consolidation. Activation is where students review previous material, take notes, learn vocabulary, and learn the steps for solving problems related to the objective or essential question. Problem solving is where students actively engage in using their new knowledge to solve problems, practice necessary skills, and put the new pieces together with prior knowledge to create meaning and understanding. Consolidation, where you take time to summarize and pull it all together succinctly, is a critical part and yet it seems to be the most often left out. An agenda can be an invaluable tool in helping you to plan effectively for your lesson summary.

DIFFERENTIATING

The goal of differentiation is to maximize student growth and success in your classroom. It can occur in both homogeneous and heterogeneous classrooms. A misconception is that only special needs students require differentiation. Differentiation simply means that you prepare instruction to meet individual student needs; however, all students have different needs at different times and respond differently to a variety of teaching styles and techniques. You should aspire to differentiate for all level of students in your classroom. Don't worry, it is not as overwhelming as it sounds.

Differentiation has become a "buzzword" in education. Many of us can attest to having participated in vague, frustrating professional development sessions that tried to reach all elementary *and* secondary teachers at the same time and used such intellectual vocabulary that the major points were lost in translation. We felt panicked until we realized that differentiation is something that most teachers already do, but don't necessarily have a name for it or include it in their lesson plans.

Differentiation doesn't have to be a burdensome task that you use only to impress an evaluator. You can differentiate in three major areas: content, process, and products. Within these areas, you can differentiate based on student readiness, interest, or learning profile. When a teacher makes modifications in one of these major areas for an individual or a small group in order to create a positive learning experience that facilitates student growth, that teacher is differentiating instruction.

Not Like That!

Mr. Zwicker assumes that all students learn the same way, require the same number of problems, and need the same level of rigor for the problems that they are presented with either in class or for homework. His focus is trying to treat everyone equally, but has not considered the idea of equity.

Ms. Wixon knows that not all students learn the same and she expends a lot of time and resources bringing students who are behind up to grade level, and in that process the silent majority (the middle-level students) get left behind. She always relegates her best students to the role of tutor. Ms. Wixon measures everyone's progress by a set standard, which is sometimes appropriate, but not always.

The end game is that we want to be sure that we are providing every student with opportunities to be challenged and to grow. Teaching, assessing, and responding the exact same way to all students will not allow you to reach all learners.

Teach Like This!

Let's begin by focusing on three classroom areas in which teachers can differentiate based on student readiness, interest, or learning profile.

Content refers to the standards, learning objectives, and curriculum. All students in the classroom may be required to know certain number facts, material, and skills. What will be most likely to change with differentiation is how students access that knowledge and the depth to which they learn it. Some of the ways a teacher might differentiate access to content include adjusting the degree of difficulty or level of scaffolding and using flexible grouping.

Adjusting the degree of difficulty of a task helps you provide an appropriate challenge for all students. You can give hints or guide lower-level students to the appropriate place in their notes. You might ask an accelerated learner to include a justification, ask a "what if this were changed" type of question, ask them to create a problem of their own, or assign a "challenge" problem. The key is to know that you don't have to create a whole different assignment if you have prepared in advance to ask them a question that requires higher order thinking skills. One of the few benefits of textbooks is that they tend to include some nice application problems, challenge problems, and ACT practice problems that you can use to benefit the students who require more of a challenge.

To adjust the level of scaffolding, you can increase or decrease the amount of peer tutoring, manipulatives, notes, or teacher provided help depending on student readiness. It is a beautiful thing to watch an at-risk student blossom. You start by providing a strong scaffold to help them achieve success. The more confident they become the less scaffolding you provide until they are able to do the majority of work independently. The irony is that high-achieving students can require as much (or more) assurance than at-risk students. When doing independent practice, make answers and examples of how the problems are supposed to be worked out readily available to students.

A third way to differentiate content is to use flexible grouping. It is suggested to plan student working arrangements that will vary widely and purposefully over a relatively short period of time. For example, when you use peer tutoring as an instructional tool for relatively new material, create heterogeneous groups with the low-middle-middle-high structure. When you have a tough objective that requires you to reteach, use a homogeneous grouping that will allow you to focus on the group that needs reteaching, give the middle level time to practice independently, and have the high level produce a product or work challenge problems.

Process refers to the activities in which the student engages in order to make sense of or master the content. Your instructional strategies are the tools of your trade. Teachers are limited when their sole source of instruction is worksheets or a textbook. Many types of activities provide opportunities for differentiation, including group investigations, learning centers, tiered activities, and alternate forms of assessments. The critical part is to know your students' readiness, interests, and learning profile. You want to make sure that you are using a variety of modalities such as visual, auditory, and kinesthetic teaching strategies to meet the needs of all your students.

Play a relay game—you can structure a group activity that includes a combination of different levels of problems that everyone can participate in without feeling the pressure of being graded. Put students in groups of four (heterogeneous grouping) and give them four problems at varying levels of difficulty. Let the group decide who is going to do each problem. Each student works one problem and checks the answer of one group member.

Do a scavenger hunt—place problems around the room that have the answer to another problem in the top right hand corner of the paper. Have groups all start at a different problem and when they get their answer they look around the room until they find their answer in the top right hand corner. When they find it, they work the problem on that sheet and so on until they have completed all the problems. You can differentiate by allowing or not allowing certain groups or individuals to use calculators. When creating the activity, you can go from easy to hard and require some students to complete only the first five of ten. You can also require advanced groups to complete this task under a time limit.

Products are items that demonstrate what a student has learned. This is probably the easiest thing to differentiate. A good product requires that students extend their understanding, apply what they know, and use critical thinking. A product can be many things: a comprehensive paper-pencil test, a project, a portfolio, a math journal, or an exhibition (verbal or visual) of a complex solution. You can differentiate by helping students to choose an appropriate product, providing assignments that vary in difficulty according their readiness, using a variety of assessments, or allowing test or quiz corrections with varying degrees of scaffolding.

As you are considering how to include differentiation within your content, processes, and products it is imperative to understand that students have different degrees of readiness, extremely varied interests, and learning strategies shaped by their experiences, resources, culture, and preferred method of learning. The ultimate goal is to make each student feel challenged most of the time. It can be a lot to think about and differentiation requires ongoing and effective formative and summative assessment. However, it is one of the best ways to go from being a good teacher to a great teacher

BUILDING CONCEPTUAL UNDERSTANDING

Building conceptual understanding means knowing more than isolated facts, procedures, or algorithms. To refine that definition a little further, conceptual understanding means the ability to comprehend mathematical concepts, operations, and relations and then apply them.

Over the last forty years of math education, the pendulum has constantly swung between more emphasis on conceptual understanding and more emphasis on procedural knowledge. Having procedural knowledge means having the knowledge of the rules and algorithms related to mathematics. This is also known as "rote" learning. Although both skills are necessary for a well-rounded student of mathematics, it is critical for teachers to ensure that students build conceptual understanding in conjunction with procedural knowledge.

Not Like That!

Mr. Walsh believes that the most important thing in his math class is the ability to memorize and complete the steps to calculate solutions. Students memorize the quadratic formula and then consistently substitute in different values to the formula and calculate by hand the solution. For many, they have become experts in the procedure, but without any knowledge for when and how this is to be applied in context. Some students do not have the ability to complete this multistep process and advanced calculations and are angry and frustrated.

Some inherent issues with only focusing on procedural knowledge include:

- It is boring and not joyful.
- It can easily be calculated using technology.
- It does not offer students the opportunity to recall in a meaningful manner.
- It represents very low level thinking on Bloom's taxonomy.
- It does not help students make connections or relationships with mathematics.

Now, let's look at an easy example that almost everyone will be familiar with: Order of Operations or PEMDAS as it is commonly referred to in classes. It is a cute trick-of-the-trade acronym that middle and high school teachers love to use to help students recall the steps for simplifying expressions. However, it is not without its problems.

Although we teach it using the phrase "Please Excuse My Dear Aunt Sally" it still just means Parentheses, Exponents, Multiplication, Division, Addition, Subtraction—a phrase that doesn't *mean* anything. After the first push of learning Order of Operations, teachers tend to allow students to use a calculator to solve the problems because it is no longer part of their standards. Students remember the acronym but they don't remember that multiplication and division are done in order from left to right, and so division might be done before multiplication in some problems. The same is true for addition and subtraction. Finally, PEMDAS does not give students the ability to recall this in a meaningful manner, thus generating opportunities for error and misconception.

The same can be said for the acronym FOIL as related to the distributive property. FOIL stands for First, Outside, Inside, Last. When multiplying binomials, it works like a charm.

However when students try to take this idea of FOILing and apply it to a trinomial, it will no longer work as the middle term is ignored, or not distributed. This quick method for remembering binomial distribution has negatively impacted future learning and students will not have grasped the distributive property.

First Terms ⟶ $(2x+3)(x-2) = 2x(x) = 2x^2$

Outer Terms ⟶ $(2x+3)(x-2) = 2x(-2) = -4x$

Inside Terms ⟶ $(2x+3)(x-2) = 3(x) = 3x$

Last Terms ⟶ $(2x+3)(x-2) = 3(-2) = -6$

This results in:

$2x^2 -4x +3x -6$

Combine like terms: $(-4x + 3x = -x)$

$2x^2 - x - 6$ This is the final answer.

Figure 2.1.   FOILing with binomials. Karin Hutchinson. Algebra-class.com.

Teach Like This!

Helping your students gain conceptual understanding requires some additional work on your part as a teacher. You have to reflect on your practices and learn how to use multirepresentations, manipulatives, and mental math, such as chunking, to go beyond procedural knowledge. When teaching a topic, it is important to share, demonstrate, and discuss how the mathematics work and why you follow the procedures that you do. Examples of this include deriving formulas, proving formulas or equations, or even drawing pictures to show why a mathematical concept works. Teachers should begin a concept focusing on conceptual understanding and then moving to procedural fluency, which is applying the procedural knowledge accurately and efficiently.

Let's look at another example using addition in early elementary math: $36 + 38$

Procedural Knowledge
line up vertically using the "carry over" method
$$\begin{array}{r} 36 \\ +\ 38 \\ \hline 74 \end{array}$$

Conceptual Understanding
36 is $30 + 6$ and 38 is $30 + 8$ now $(30 + 30) = 60$ and $(6 + 8) = 14 = 10 + 4$
Therefore $36 + 38 = 60 + 10 + 4 = 74$

Getting students to show different representations of the same concept is important for conceptual understanding. In high school it becomes crucial for students not only to recognize different methods or strategies but also to evaluate a problem and choose the most efficient strategy.

For instance, when teaching algebra students how to solve quadratic equations, you have multiple methods (graphing, factoring, completing the square, and the quadratic formula). These can all be accomplished one at a time using procedural knowledge. However, students without conceptual understanding begin to flounder when they are given a mixture of types of problems and must choose which method would be best. They need to recognize, for example, that if the answer is going to be an irrational number or a non-integer then factoring is not ever going to work. Students with conceptual understanding can choose the best method for the problem as opposed to trying to make the problem fit a particular method.

The tricky part is to explain to parents that although you are writing out the steps, you are really preparing students to do mental math and in doing so teaching them critical thinking skills that will be applicable to the rest of their lives.

SOLVING PROBLEMS IN MULTIPLE WAYS

> It is better to solve one problem five different ways than to solve five problems one way.
> —George Polya, mathematician and father of problem-solving techniques

This is a critical issue in mathematics education, which does not devote sufficient time or emphasis to making sure that students know how to solve problems using different methods.

Not Like That!

Jeremiah is so excited to be getting his first math test back from Mrs. Booker. He feels confident in his answers and believes that he really knows the material. Excitement and joy quickly turn to anger and frustration when his test is returned with a grade in the "C" range. At the top of the page Mrs. Booker has written in red ink: "This is not the way I showed you how to work these problems."

The answers themselves are correct. The question that begs to be asked is "What exactly was being graded on the test?" Obviously it is not "Can Jeremiah solve the problem?" In actuality, Jeremiah is being graded on his ability to mimic a *method* modeled by a teacher.

By the time students get to the secondary level many have lost the initiative for true problem solving. For example, Mrs. Garland assigns a group discovery activity on dimensional analysis of units of measure with which the students would be familiar. Many groups have difficulty getting started.

One student raises her hand and asks "How do *you* want us to do this? I don't want to get it wrong." The red flag is the emphasis on the *you*. The students have thoughts on how to solve, but are concerned with what the teacher considers to be an acceptable strategy.

Avoid being the teacher who demonstrates a particular procedure for solving a problem and requires everyone to solve the problem using that same procedure or process. Simply asking students to solve twenty almost identical problems using that same approach over and over, without sharing how students could use different approaches or even create their own methods or processes for solving the problem is asking them to mimic, not problem solve.

Teach Like This!

There is great value in allowing students to explore and contrast many different ways to solve problems.

—Jon Star

The phrase "Work Smarter, Not Harder," coined by Allen F. Morgensen in the 1930s, is very appropriate here. The first step in teaching problem solving is to look for opportunities for students to have multiple entry points or strategies for solving a problem. Spend the extra time to allow them to explore their options. Take time to discuss strategic choices. For instance, when solving an equation that has parentheses, you can start with a simple problem:

$5(x+3) = 20$

Students will immediately use the distributive property. They learned to do this problem using their procedural knowledge in the eighth grade by using the distributive property. However, in Algebra I, the focus should be on conceptual understanding, so ask the students to find a "shortcut." Students love it when you give them an opportunity to "do it in their heads" instead of showing work. If they have good conceptual understanding, they will know that this is a multiplication problem that can be solved by using the inverse, division, first.

$5(x+3) = 20$
$x + 3 = 4$
$x = 1$

Then, ask them to find a strategy in this method. Eventually they will reach the understanding that not all problems are candidates for this method, such as:

$$3(x - 4) = 10$$

Why? Because we have to work with fractions too soon into the problem. In this case, the distributive property is the best method.

$$3x - 12 = 10$$
$$3x = 22$$
$$x = 22/3$$

Take an extended amount of time to examine a single problem and identify many different approaches for solving it. Some of the approaches can be shown by the teacher; however, giving students time alone, in pairs, or in small groups to find their own unique process for solving the same problem is one of the best ways to promote conceptual understanding.

For instance, ask the students to find three consecutive integers whose sum is 96. Most students will use a guess and check method and get the correct numbers. Then you can take the same problem and make it a little more difficult using consecutive even or consecutive odd integers. When students have developed a conceptual understanding of what numbers they are looking for you can introduce how to write an equation that will give them the numbers.

Another option would be to first have the students understand and apply a single approach demonstrated by the teacher. Once they attain proficiency, the teacher could ask students to individually attempt to solve the same problem creatively using a different approach.

For example, when teaching how to determine the surface area of a rectangular prism, the teacher might first discuss the properties of a rectangular prism, and then lead the class in measuring the side lengths and finding the area of all six faces. Students use a rectangular prism model on their desks to search for another way to find the surface area. Most students will realize that the opposite faces are congruent and can come up with the same answer a different way.

If you want a student to use a particular strategy or method, you should show them why. You can either show them an instance in which their method won't work or give them a sneak peek into an objective that will require them to use the one you are currently touting. When students understand the purpose they are more willing to comply with your rules.

Nationally, a staggering 20 percent of gifted students drop out before completing high school. If you have ever worked with a gifted student, then you know that their brains and thinking are very different from the average person. Gifted students see patterns and make connections in a different, but very real way. Trying to force them into a particular strategy when they have found another workable method is like trying to make a right-handed person write with their left hand—it feels unnatural.

If you cannot provide a mathematical reason as to why they shouldn't use a method different from yours, then you need to open your mind to new and sometimes exciting approaches. A gifted student might work a problem very differently from the approach planned out by the teacher, but on inspection the teacher might see that the student has come up with a better approach. Good teachers allow that experience to mold how they approach that problem, or that type of problem, in future lessons.

ACKNOWLEDGING MISTAKES

Everyone makes mistakes and sometimes they can be very embarrassing. In mathematics, mistakes can stem from a simple calculation error, a lack of understanding of the concept, or a common fallacy. How we handle the mistake when it is made determines whether the classroom is a place of shame or a place of growth and learning. Having a classroom environment that accepts mistakes will make students better problems solvers and mathematicians.

Not Like That!

Mr. Loyd is going down the row and having students to give their answers to a multistep equations worksheet. When he calls upon Kaylee she gives an incorrect answer. He sighs loudly and then calls upon the student behind her who gives the correct answer. Kaylee is embarrassed and all that she has learned is that she doesn't have the correct answer. Other students, who had the same wrong answer, are also embarrassed. Philip has missed several already and is becoming anxious knowing that he will soon be called on and he doesn't know if he has a correct answer or not. If he is wrong, how will his teacher react?

After several students have given wrong answers, Mr. Loyd vents his frustration by saying hurtful things such as "I don't understand why you kids don't get it," "Why aren't you paying closer attention?" "This isn't rocket science," "You just need to study more," or "How many times do I have to show you this?"

Ultimately, this approach results in an atmosphere where mistakes are not tolerated and growth is inhibited. No one likes to be embarrassed or belittled, so students will find ways to keep from looking inadequate, including cheating, refusing to do work, or becoming a discipline problem so they can be sent away from the classroom.

Teach Like This!

One simple strategy to alleviate shame and embarrassment about making mistakes in your classroom is to reward students for catching you when you make a mistake. It is very easy to make mistakes when you are computing, writing, thinking about what you are going to say (or do) next, and thinking about how you can generate critical thinking questions for your students all at the same time. Laugh at yourself and say "everyone makes mistakes, but not everyone can catch a mistake! So when you catch me (the supposed expert) in a mistake I am going to reward you for your creative thinking."

Let's go back to Mr. Loyd and his classroom where the students are learning multistep equations and see how a mistake can be turned into a positive growth experience. When Kaylee gave a wrong answer, Mr. Loyd could have said, "Kaylee, if you made a mistake on this problem, more than likely someone else made the same or a similar mistake. Do you mind if I show your work to the class so that we can use this as a teachable moment that we can all learn from?" Mr. Loyd projects the problem and then asks the class to silently look at it and see if they can determine where the mistake was made.

Before allowing anyone to say anything he gives Kaylee the opportunity to see if she can see her own mistake. If she cannot find it, he gives others in the class the opportunity. When a consensus and correct answer have been reached he thanks Kaylee for allowing him to use her mistake as an opportunity for everyone to grow. He reminds everyone that finding mistakes is sometimes harder than working the problem itself.

At the beginning of the year, reiterate frequently that it is important to not only be able to get a correct answer but to be able to analyze a problem and "fix it," which is truly a real world application. If students (and teachers) didn't make mistakes then the class would lose the opportunity to stretch their minds and develop true problem solving strategies.

Let's look at one more example. Ms. Sherrod is teaching a lesson on exponent rules. During guided practice she calls on Isabella to answer the problem $(3x^2)(2x^3)$. Isabella answers incorrectly by saying "$5x^5$." Ms. Sherrod responds by saying "Isabella, although your answer is incorrect, I am so glad you said that. I forgot to remind everyone about a common mistake when multiplying monomials. You have to remember to multiply the coefficients (the numbers in front) but add the exponents. So you have to be sure that you are identifying coefficients and the operation needed to simplify that piece as opposed to the process needed to simplify the exponents."

"So, Isabella, if I give you a similar problem like $(4x^3)(2x^2)$, describe how you would find the coefficient. Now describe how you would find the exponent for the variable." Isabella gives correct answers and Ms. Sherrod says "Awesome, I was confident that you knew it and thank you for reminding me about this common mistake."

Knowledge of your students is also a crucial part of how you acknowledge and correct mistakes. Your students come with a wide variety of personalities and levels of confidence; use this to your advantage. For shy or weak students, for instance, you could walk by their desk and check their answer to a particular problem. If it is correct be sure to call on that student later to answer that problem. When you are facilitating and you see where a student makes a mistake, ask them if you can use their mistake as an example of a common error for the class.

Stanford professors Carol Dweck and Jo Boaler are experts in the benefits of students having a growth mindset as compared to a fixed mindset. Dr. Dweck has publicly stated that "Those with a growth mindset believe that their intelligence can be developed, while a fixed mindset reflects belief that intelligence is unchanging." There are many great books written by Dweck and Boaler solely on this topic such as *Mathematical Mindsets* and *Mindset: The New Psychology of Success*. One key point about growth mindset is that teachers accept student mistakes and students know that making mistakes is more than acceptable; it is part of the learning process.

Being able to make and then correct mistakes benefits students. In a classroom in which students are free to make mistakes, they don't get embarrassed or ashamed, the teacher acknowledges the mistake in a positive manner and helps guide students toward the correct process, and everyone in the class benefits from analyzing a mistake.

COMMUNICATING WITH STUDENTS

Communication is a critical component of developing a mathematically literate student, and should be a specific component of the lesson each day. Both the student and teacher benefit from effective communication. First of all communication strengthens the connection between teacher and student. It also helps students to achieve their goals, creates opportunities to expand student learning, and provides teachers with insight into student and classroom understanding of an objective.

Not Like That!

In Mrs. Jones's eighth-grade class, students communicate infrequently during class, and then only as short responses to direct questions from the teacher. Common responses are 86, $x = 4$, or false. The only dialogue between teacher and student comes when Mrs. Jones tells the student if the answer is correct or incorrect. These types of answers are not indicative of true understanding of a topic. Students are not developing or communicating their ideas and processes related to the mathematics concept, and the teacher cannot assess the students' level of understanding.

Teach Like This!

Have you ever had that student who starts yelling, "I have the answer!" as soon as she finishes a problem and has a solution? At the beginning of each new school year you can use this moment as a platform for sharing with your students how you plan to use communication in the classroom. Ask the student, "Are you sure you have the answer?" They always emphatically say "Yes!" Then ask, "What is the first step you used to solve this problem?" This typically produces with a look of confusion and panic. However, it sets the precedent for how the class will be required to communicate with the teacher, and also enhances students' listening skills.

This process of communication also allows you to differentiate in the classroom. As your knowledge of your students increases you can use communication skills to guide students to higher order thinking. The complexity of the communication should reflect the student's current level of understanding.

Communication can occur in many ways, including: talking to peers, talking with class, writing, explaining thinking, and explaining through multiple means such as pictures, diagrams, sketch notes, Venn diagrams, graphic organizers, Cornell Note-taking guides, and interactive math notebooks. Communicating our thoughts can make us slow down and more carefully think through each step of the process of mathematical problem solving.

It is critically important that we guide students in the communication of problem solving. If we only allow students to mimic a problem that we have solved over and over again they may get a correct answer but they will tend to make more mistakes during the process. If they don't understand the solution, they can't apply it to a related task.

However, when we focus on the "what" and the "why" by communicating our thought process and organizing information in oral and written formats, we can solve problems more efficiently and can transfer that knowledge to a new but similar problem. As teachers, we can guide the discussion determining the "what" and "why" and "how" by asking questions such as:

- How did you decide where to start?
- What made you group these items together?
- What do these two figures have in common?
- Could you have worked this a different way and still gotten the same answer?
- Did you use the most effective strategy?
- Why did you choose that strategy?

- Is your answer reasonable?
- If you do this . . . then what is going to happen?
- What is the key thing to remember from this definition?
- Write a couple of sentences comparing and contrasting these.
- What property or definition allows you to make that choice?

The more specific we are in our questioning the more specific the students learn to be in formulating and communicating their processes. We help them to adopt a higher level thinking when we teach them to focus on the what, why, and how of the problem-solving process. When we simply tell students to "go at it" they make lots of mistakes and errors and may eventually find a solution, but the learning gets lost in the struggle. Trial and error is a good problem-solving skill if there is conceptual understanding behind what is being asked and what are appropriate values to test out, but when it occurs without a foundation, students likely will not have a reasonable starting place to approach the problem.

It is the role of the teacher to allocate time for these different modes of communication and teach the students how to communicate in each of these modes. Students often prefer a certain type of representation or approach to solving a problem, and many students will also have a preference for how they want to communicate their reasoning in math. Teachers should be flexible and accommodating to the style in which students want to express their reasoning and understanding of the mathematical concept.

DOING MENTAL MATH

Mental math is the ability to do computations in the mind without the aid of paper and pencil or calculator. Of course, there are lots of "tricks" out there that will help students solve problems mentally as they become more mathematically proficient.

Not Like That!

Ms. Blakely's fifth-grade students are presented with the task of finding the area of a rectangle with a base of 102 feet and a height of 13 feet. Most students begin writing the standard algorithm while some take out their calculators to compute as the pleased teacher watches. One student immediately raises his hand to declare that he knows the correct answer of the area to be 1,326 square feet. Ms. Blakely asks to see his work justifying the answer, and the student explains that he was able to calculate it mentally by breaking up the 102 into 100 + 2 and easily calculate the answer. Instead of being applauded for being able to calculate the answer mentally using a creative approach, the teacher dismisses the answer since work must always be shown to earn credit.

Teach Like This!

If you are not sure that the student was actually able to do the computation, ask them (in a praising manner) to explain their process. A typical rule of thumb is that if you can do it in your head then there are probably students who are also capable of doing it in their head. The surprising part is that it will not always be your "bright" students who can do this. Lazy students are quite creative when it comes to math and they would prefer to do it in their heads as opposed to showing work. There is a time to show work, but there should also be a time for encouraging students to use logic and mental math.

Mental math usually involves three basic processes:

- Being able to use chunking to break larger numbers into smaller parts more amenable to mental math skills. For example $232 + 45 = 200 + 30 + 40 + 2 + 5 = 277$.
- Being able to recognize when numbers have properties that will allow you to use a pre-worked shortcut. For example factor $25x^2 - 49 = (5x - 7)(5x + 7)$.
- Being able to rearrange a problem in order to make calculations simple. For example find the area of the triangle given the formula $A = (1/2)(7)(6)$. Rearranged $A = (1/2)(6)(7)$ results in 21.

Being able to estimate and compute mentally is a critical skill that is quite often ignored or discouraged. Having mental math skills builds conceptual understanding as opposed to just memorization. It also develops a strong number sense which is lacking in many of our students.

Let's look at a simple example. When adding the numbers $17 + 24$ students are taught to set the problem up vertically and then to use a series of steps including "carrying." This process almost always requires that a student write the problem out in order to successfully complete the steps without confusion. However, if a student has a strong conceptual understanding they recognize and can chunk 17 as $10 + 7$ and 24 as $20 + 4$ then $10 + 20 = 30$ and $7 + 4 = 11$ therefore the sum is 41. This strong number sense doesn't rely on an algorithm that students must memorize.

Mental math is also logical. If you look at the example above, when it is taught to students using the standard algorithm, students must work from right to left. We read from left to right so right to left can feel awkward. Teaching students to use chunking allows them to work from left to right in a more natural flow.

When students feel comfortable using mental math it can help to put them in the habit of checking the reasonableness of an answer. All math teachers lament the fact that students won't check their answers. However, if we encourage them to use mental math, they are more likely to check and be able to determine whether the answer is reasonable.

For example, when working with a quadratic equation word problem that asks for the time it takes for a ball to land on the ground, students get both a negative and a positive answer. If they have been taught to think and not to simply plug answers back into the equation, they will realize that time cannot be negative therefore the negative answer is irrelevant in the context of the problem and the correct answer is the positive answer.

When approached in an appropriate manner, mental math can be fun and has many, many benefits to the student. If you present it as a shortcut students will be all over it and love to compete with other students to see how fast they can solve a problem mentally. This is important as students begin to prepare for SAT, ACT, or any other type of timed testing. The ACT math portion requires students to solve sixty problems in sixty minutes. When we require students to keep showing work for concepts that they mastered several years ago, we are handicapping them and eroding their confidence.

We should encourage students to build their ability to do mental math because:

- It keeps the brain sharp and builds confidence in their skills.
- It allows students to judge the reasonableness of an answer.
- It encourages students to become efficient problem solvers.

They are not waiting on a formula or being told how to solve the problem. The biggest complaint in industry regarding education is that we have not taught students to be problem solvers. We have taught them to follow algorithms that are not always present in the workplace.

- It actively engages students in the process of mathematical thinking.
- It allows students to use differentiated strategies in problem solving.

Mental math activities can occur during the warm-up phase of the lesson or students can be encouraged to attempt to solve computations mentally when appropriate. It is vital for students' future success that teachers model and actively engage students in practicing and reinforcing mental math skills.

A quick example of how to model mental math is teaching students how to add integers by identifying zero pairs.

$-5x + 8x$
$= -5x + (5x + 3x)$ decompose $8x$ as $5x + 3x$

$$= (-5x + 5x) + 3x \text{ regroup and recognize } -5x + 5x \text{ as } 0$$
$$= 3x$$

A good place to begin practicing and reinforcing mental math skills is on objectives that students have already mastered. For instance, students learn to solve one- and two-step equations in middle school. At the high school level you can practice solving those types of problems mentally. For example:

Solve $2x - 7 = 9$
Review the steps verbally: step one is to add 7, step two is to divide by 2
Solve mentally $9 + 7 = 16$ and $16/2 = 8$ therefore $x = 8$
Check mentally $2(8) = 16$ and $16 - 7 = 9$

Writing it all out is a time-consuming task and students don't buy into it. The older they get the more resentful they become of not being allowed to save time by using mental math and not having their strategies recognized as valid problem-solving methods. Teaching them to use mental math will be a winner not only for them, but for you as a teacher as they gain conceptual understanding, which ultimately contributes to long term retention.

KEY TAKEAWAY POINTS

- Creating and sharing objectives and essential questions at the beginning of the lesson should become the norm of all teachers. Reflecting on these at the summary phase of the lesson provides a check for understanding as well as a closure for the lesson.
- Posting and discussing an agenda for your students benefits everyone in the class. We expect an agenda for our meetings and classes and students do as well.
- Having warm-up problems to refresh prior knowledge or concepts from the previous day are an excellent way to start each class period.
- Five to seven minutes should be saved at the end of every lesson to summarize and assess the learning while also previewing the homework for the next class period.
- The ultimate goal of any mathematics teacher should be the build conceptual understanding of the topics covered. This will help students recall the concepts long-term.
- Students should be encouraged to solve the same problem more than one way. Solving a problem only using the method modeled by the teacher does not deepen understanding or the core mathematical concept.
- We want our students to show their work so that we can try and understand their thinking and identify mistakes, but there is much value to teaching and practicing mental math. It should be encouraged.

Chapter Three

Assessments

Throughout the week of instruction, Ms. Baker conducts frequent, quick assessments during lessons to determine whether her students are learning the concepts related to the objectives of the lesson. These quick check-ins also identify the students who are struggling and need additional support before the formal assessment. The formal assessment uses a variety of question types to identify what the students know and what they don't know. For the open response questions, Ms. Baker develops a rubric and shares it with the students so that the expectations are clear. She establishes proficiency levels for the assessment based on standards for expectations and makes modifications for students with disabilities.

After correcting the assessment, Ms. Baker analyzes each question to determine success rates and identify any common misconceptions. If they attend a voluntary review of the material, she offers students the opportunity to earn back part or all of the points they lost on the original assessment. Her ultimate goal in the classroom is that learning take place.

Assessments are a critical piece of being an effective teacher. You must assess student learning in a meaningful and authentic way. You must know a variety of quick formative assessments that you can use throughout a lesson and understand how to tell what changes you should make based on what you learn. You must know what types of interim assessments to give at the end of a unit and also how to *analyze* the results and use those results to guide instruction.

You should also know about summative assessments and the impact that they should have on your long-term instruction and goals as an individual teacher and as a member of a team of teachers. Finally, you need to consider allowing students to retake assessments under the correct circumstances.

This chapter will detail ways to effectively create, modify, and use assessments to improve student learning.

FORMATIVE ASSESSMENTS

Formative assessments are designed to check students' immediate understanding in the moment during the lesson. Examples of formative assessments include:

- questions students can answer verbally or by using clickers or other technology;
- student self-assessments (such as thumbs up, sideways, down, or 1–5 scale);
- exit tickets; and
- warm-up problems.

Not Like That!

Mr. Geunith completes his sixty-minute lesson on factoring trinomials and believes that the students understand the process. However this was not the case as he did not assess the learning. Instead of moving on to the next lesson, factoring trinomials with an "a" value greater than 1, he now has to quickly try to adjust and reteach the concept that he had wrongly believed the students understood.

Have you ever given a quiz or test and then been extremely disappointed by the grades your students earned? You realize suddenly that the students didn't understand the concept as well as you had thought, and then start going through a mental checklist: *yes, I did teach the material; yes, I assigned homework and the students completed the assignment. Where did things go wrong?*

Mrs. Counce has given a homework assignment and diligently checked all the answers. She feels confident giving a quiz, but the quiz scores are abysmal. What went wrong? When you use homework to check for understanding, you have no way of knowing if (1) the student completed the work themselves; (2) if the parent helped the child; (3) if the student used technology, such as an app that lets you take a picture of a problem and generates a solution; or (4) if they copied from a peer. Therefore, using homework as a formative assessment technique can be very misleading if you want to gauge actual student understanding.

Teach Like This!

Ms. Reynolds plans and implements multiple formative assessments throughout the lesson. She does this to see whether the class as a whole understands the objective and to identify individual students who might be struggling.

Her first formative assessment occurs when she gives her warm-up problem to the class. She quickly identifies three students without the prior knowledge necessary of the current concept. During the guided practice portion of the lesson, Ms. Reynolds asks the students to give her a "thumbs up" or a "thumbs down" if they would like for her to provide another example problem. Notice that she avoids asking them to use the "thumbs up/thumbs down" method to ask them if they *understand* the material. Students are typically too embarrassed to say that they don't understand, so the alternate phrasing tends to more accurately gauge student understanding. Be careful of using methods that ask students to relate their own level of understanding. Most students do not realize when they are wrong.

After she launches the lesson and students begin the exploration phase, she pulls those three students aside to review the skill that that they lack, so that those students are able to acclimate to their partner work as the lesson continues. In order to keep from slowing down the class to wait for those three students to finish, she differentiates by allowing them to complete fewer problems and chooses those that are the most critical for them to understand to gain mastery of the objective.

By circulating through the classroom and observing student work, Ms. Reynolds is also able to do a second assessment of the students' understanding of this new concept and offer support for those who are struggling. If a significant portion of the class is unable to complete the activity, Ms. Reynolds can immediately pull the class back together to address these struggles or misconceptions instead of waiting for a more formal assessment.

Before solidifying the homework problems for the evening, Ms. Reynolds takes two minutes during the summary phase of the lesson to ask guiding questions. Students answer anonymously using a classroom set of clickers or a variety of different apps. This helps her choose the appropriate homework for the class and modify the assignment for students who are struggling or have already mastered the concept.

Even if you or your district lacks technology, there are many ways to do formative assessment without it. Exit tickets use writing prompts to allow students to show mastery and understanding of the objective. When you are ready to go "next level" with students, use short writing prompts that ask students to explain their thought process, such as: *explain the first step in solving this problem; define in your own words what an equilateral triangle is; tell me the hardest thing for you today; explain the difference between a rectangle and a square;* or *describe how multiply two fractions.*

Another good opportunity for formative assessment is at the end of the guided practice portion of the lesson. Mr. Nathan asks his students to work one to three problems and bring them to him to check. This gives him an opportunity to not only check their process, but also to check that they have taken the notes that he has given and that they have worked the example problems.

If students have accurately worked the problem(s), have them begin an independent assignment. If students have made mathematical errors but understand the underlying concept, either help them to find the mistake or give them a hint as to where they should look for it. You can put students who are struggling to understand the concept together in a small group that you facilitate.

A wise teacher once said, "Understanding the purpose and utilizing formative assessments has been a lifesaver for me in the classroom. I no longer have to spend copious amounts of time grading papers, creating additional materials, and using needed classroom time for re-teaching and I am not blind-sided on test or quiz scores."

USING INTERIM ASSESSMENTS

Interim assessments, also known as benchmarks, fall between the short informal formative assessments that drive immediate classroom instruction and the summative assessments at the end of a grading period or a course. Teachers and schools can use these assessments on a regular basis to monitor the progress of student learning and gauge development of mathematical concepts. They provide formal data prior to the end-of-year summative assessments or standardized tests.

Not Like That!

Mr. Jensen complains when the quarterly half-period assessments are scheduled, as he believes that his own end-of-chapter tests give him enough information to monitor student learning. He resents having to use common assessments because he has used the same material for years and his students perform well on his chapter tests, and these assessments take away from his precious instruction time. However, when his students take the new state-mandated standardized test, they do not perform well.

Mr. Jensen administers the assessment as directed by his data team; however, he does not take time to analyze, interpret, and reflect upon the results. Nor does he plan his scope and sequence to coincide with the material that will be assessed on the benchmark. He has his own timeline for when and how long to teach each objective. He also has his favorite activities and objectives even though they are no longer part of the standards for his grade and are assessed at an earlier grade level.

Teach Like This!

Although it's a valid argument that we are over-testing our students, giving short, quarterly, common interim assessments can provide all teachers with valuable data. These data can be used to help guide full class instruction and intervention.

For instance, Miss Fiore is given a directive from her district to administer interim assessments. She collaborates with her math department and her intervention coordinator to develop interim assessments that are aligned with the state standards and a pacing guide that will allow her class to cover all the objectives in a timely manner for state standardized testing. She understands what is being assessed, and her team has developed a plan for students who are underperforming and will need some type of remediation or intervention.

Miss Fiore has taken time to learn the testing tool and has collaborated with her peers to be sure she understands what is being assessed from her students and what the results will tell her. Once she has the results of the assessment, she spends time analyzing, interpreting, and reflecting on the results, and uses this to guide her instruction and possible interventions for her students.

With many states now requiring some type of Response to Intervention (RTI) built into the school day, it becomes even more critical that teachers, departments, intervention coordinators, curriculum supervisors, and districts develop interim assessments to determine student baselines and achievement growth. Technology has made it easier than ever to use all these data points to truly track a student's progress and provide interventions as they are needed, so that students don't get lost and systematically passed on to the next grade. Just imagine if we were able to identify these students earlier and provide them assistance *before* it becomes an overwhelming and impossible task! Quality interim assessments can make this a possibility when used in conjunction with an intervention program.

Not only can interim assessments benefit students who are behind, but they can also be used to identify students who would benefit from more challenging and enriching material—not just gifted students, but high-achieving students who can do self-directed work.

Interim assessments are designed for three main purposes: instructional, evaluative, and predictive. They should provide useful information to the classroom teacher, the math department, parents, school administrators, and the district's curriculum supervisors.

Let's look at how the interim assessments can be used at each level.

1. In the short term, they can provide results that help teachers adapt instruction and curriculum to better meet student needs, including re-teaching, tutoring, and remediation. They can allow us to see which objectives students are having difficulty achieving or retaining, so that we spiral in those objectives with future teaching and assessments.

2. In the long term, they should drive decisions regarding curriculum and the provision of human and other resources. As education changes, we need new resources and materials. Textbooks are no longer the "go to" resource, particularly with so many school districts achieving or work-ing toward a one-to-one technology resource for students. Interim as-sessments allow us to not only see flaws in our teaching but in our resources.

3. They can be used to inform parents of a student's progress toward mastery of specific standards and objectives. When parents understand the specific area in which a student needs help, they can help their child by using internet resources such as videos or hiring a tutor.

4. They help administrators determine learning patterns for large groups of students and use that data to drive curriculum and professional development decisions.

5. They can be used to place students in the appropriate intervention, such as basic skills intervention or standards remediation, on a very particular objective, or to give them access to enrichment opportu-nities, such as gifted programs.

6. They can be used for predictive purposes to determine each student's likelihood of meeting the criteria on standardized and end-of-course testing.

7. They can be used to document milestones and to determine if students are on track for meeting grade-level expectations.

8. They provide consistency for students across the school and district ensuring that all students are receiving quality instruction.

Summative assessments and formative assessments are valuable, but as dis-cussed above, educators need additional data such as interim assessments to track student learning and areas of need throughout the course in order to improve student performance. Ultimately, interim tests help teachers better understand what students do or do not know and what objectives and con-cepts they must focus on to enhance grade level performance. The data derived from the interim tests can be a powerful tool when used correctly at the classroom, school, and district level.

UNDERSTANDING SUMMATIVE ASSESSMENTS

Let's start by clarifying the difference between formative, interim, and summative assessments. Formative assessments provide feedback as students are engaged in learning single objectives. They allow teachers to improve their teaching and identify the strengths and weaknesses of students in targeted areas. Students' needs are addressed immediately and the formative assessments are low stakes, meaning that they have little or no point value and are generally used for instructional purposes only.

Interim tests are more formal than formative assessments. They are a type of summative assessment, but they are typically a shorter version, not as comprehensive, and not as high stakes as most summative assessments. Summative assessments often correspond to students' grades, while interim assessments usually just measure progress over the course of the academic year.

Summative assessments usually pertain to the outcome of a specific program. Although summative assessments can be used to determine individual student learning, class performance, and progress on particular standards, they are more typically a way to assess student learning at the end point of a unit, semester, or year.

Summative assessments can be standardized tests or have a high point value or consequential value for a student's grade. Classic examples of summative assessments include state-mandated end-of-course tests, district benchmarks, chapter and unit tests, or semester exams. The scores can be used for accountability for schools (AYP) and for students in the form of report card grades. They can also be used for placement of students into different levels of mathematics classes.

Not Like That!

Mr. Plant does not believe in integrating the standards or types of questions that are on his state's end-of-year standardized test because he believes "teaching to the test" is not beneficial to his students and thinks he should decide which topics and style of questions should be presented to his students. This is a problem when his students move to the next level of mathematics and there are gaps in their learning that another teacher will now have to fill in before they can meet their own standards.

The old saying that "it takes a village to raise a child" is just as true in education. Because math continually builds upon itself, it takes everyone from kindergarten on up to present the students with a cohesive, consistent, and gap-free mathematics curriculum.

At the other extreme is Mrs. Johnson, who stops regular instruction two months before the standardized test to give students practice tests and hints to answer multiple choice questions. Being drilled for weeks in preparation for this high stakes test does not engage the students. There is a time and place for review, practice tests, and test taking tips—but they need to be integrated in with active instruction.

Mrs. Tittle teaches up until the time of the test, but once the state-mandated tests are over, she stops teaching and allows the students to spend their time playing or doing frivolous activities. She doesn't go back to focus on objectives that were only touched on because they were not going to be tested or work on the skills that will benefit students the most at the next level of mathematics. Teachers at the middle and high school level can use the time after standardized tests to prepare students for even more high stakes tests like ACT and SAT. For instance, Mrs. Moore, a ninth-grade Algebra I teacher, spends time after her state-mandated test teaching objectives that will be assessed on the ACT, or in Geometry or Algebra II classes such as right triangle trigonometry, matrix operations, and geometric formulas.

Teach Like This!

For better or for worse, high stakes standardized tests are here for the foreseeable future. Instruction throughout the year should be focused on the standards and these same standards will be assessed at the district, state, and federal level. Knowing which standards are more highly assessed can play a role in determining your pacing guide and points of emphasis.

Mr. Ortiz has met with his Professional Learning Community and they have researched which standards have to be taught and which standards will be counted as major content and supporting content. They have created common interim and summative assessments that utilize a variety of question types including multiple choice, short answer, multiple select, true/false, matching, extended writing, and performance tasks. They create each summative assessment to contain spiraled objectives from the major content that so that students utilize prior knowledge and activate their recall and retention. This shortens the amount of time that he will have to spend reviewing for the end-of-year test.

He gives a comprehensive summative assessment at the end of each nine weeks. It counts for 25 percent of a student's final nine weeks grade, which is consistent with the school district's and state's policy that the last nine weeks end-of-course grade will automatically count as 25 percent of the student's grade. This also allows him to identify students who struggle with retention or who may have test anxiety so he can help them develop strategies for taking high stakes tests.

It is not the design of the test that makes it summative, but the way the results are used. A key component of summative tests is that they are designed to be *evaluative* rather than diagnostic. They can be used to evaluate student learning and academic achievement at the conclusion of a predetermined instruction time frame.

AUTHENTIC ASSESSMENTS

On a more macro level, assessment can be based on presenting students with "engaging and worthy problems or questions of importance, in which students must use knowledge to fashion performances effectively and creatively. The tasks are either replicas of or analogous to the kinds of problems faced by adult citizens and consumers or professionals in the field" (Wiggins, 1993, p. 229).

Authentic assessments measure significant and meaningful accomplishments. They actively engage a student's voice. As opposed to multiple choice answers or standardized tests, they measure students' understanding of a concept in a meaningful format, and show what students know as opposed to what they do not know. Authentic assessments often use a rubric to measure learning. Some examples might include a portfolio, presentation, demonstration, or project, which can all be done in an individual or collaborative manner.

Authentic assessments can also be performance assessments or performance tasks which ask students to engage meaningfully and perform real world tasks which call on them to demonstrate essential knowledge, skills, and applications. Authentic assessments are occasionally referred to as "alternate assessment" because it has the student test in a nontraditional format. Alternate assessments, unlike authentic assessments, do not necessarily ask students to engage in real world application problems.

Not Like That!

Ms. Jones believes in a traditional classroom. Every end-of-chapter test in Ms. Jones's class is identical. The questions have forced-choice answers such as multiple-choice, short answer, true/false, and or matching and maybe one or two open-ended questions. This type of questioning mirrors the state's summative end-of-year test, and does not offer the opportunity for students to demonstrate the knowledge they created during the unit or chapter. Students need only to recall information in order to complete the assessment. Each answer is either correct or incorrect and is very difficult to identify where student misconceptions occurred.

The philosophy behind this traditional method is that it is a school's mission to provide students with a certain body of knowledge and skills. Therefore, schools must teach this body of skills and knowledge. Then students are tested on this knowledge. In other words, these are the standards that make up the major content and this is how they are tested. The standards drive the curriculum and the assessment to determine if the students have mastered the state objectives.

Teach Like This!

Mrs. Hirsch offers a variety of different types of assessments throughout a unit or term, including projects, presentations, and performance tasks. Just as each student learns in different ways, Mrs. Hirsch realizes that students can demonstrate their learning in different ways. All students keep a portfolio of their work that includes their interactive math notebook and performance tasks (authentic assessments), foldables, notes, and math journals. When students complete an authentic assessment task, they first self-assess it using an agreed-upon rubric, and then include it in their portfolios.

Authentic tasks do not necessarily have to be long drawn-out activities or projects. An authentic assessment simply asks students to perform a meaningful task that will link to the real world. The assessment determines the skills and knowledge required for students to perform in a meaningful way that will replicate a real world challenge. This is sometimes referred to as planning backwards.

The following is an example of a simple performance task and the rubric that guides it.

> *Performance Assessment*: The new indoor race track offers a frequent racer rewards card whereby the charge is $20 per race and after every 10 races you get a free race. Races do not have to be on the same day.

Table 3.1.   Sample Performance Tasks

1) Create a table that shows the relation between the number of races and the cost for up to ten races.

2) List the ordered pairs so that the number of races is the domain.

3) Identify the range of the relation.

4) Graph the relation.

5) Write three questions that you could ask a classmate about information that could be obtained from this relation/graph.

6) Identify the inverse of the relation using ordered pairs.

7) Graph the inverse of the relation.

Table 3.2. Sample Performance Rubric

DOMAIN, RANGE, and INVERSE	RUBRIC
The core objectives A) student will be able to identify domain and range B) student will be able to illustrate a relation using multiple representations: ordered pairs, table, and graph C) student will be able to illustrate the inverse of a relation Points possible are indicated in ( ).	**Points Earned**
1) Table has correct values (1) Table has been labeled correctly with x and y (1)	2
2) Ordered pairs are written with the x values as the domain (1)	1
3) Student identified the range (1) Student wrote the range using set notation (1)	2
4) Student graphed the relation but did not label axis or scale (1) **OR** Student graphed the relation and labeled scale (2) **OR** Student graphed the relation, labeled scale, and labeled x- and y-axis (3)	3
5) Examples (1 point for each reasonable question) a) If Jason races five times how much is his total cost? b) If Kim paid $90 how many times did she race? c) How much will Toby pay for eleven races?	3
6) Student identified the inverse using either a table or a set of ordered pairs (1)	1

Don't feel forced to be in one camp or the other regarding authentic assessment. A teacher needs to balance curriculum and assessment and use both traditional and nontraditional means in order to meet the needs of students and make sure that they are ready for college and careers.

With traditional assessment we are told not to teach to the test. This makes everyone feel like we are playing a "surprise" or "gotcha" game. Students and teachers feel inadequate when trying to guess what is on the assessment and how it will be assessed. It is also easy to cheat on traditional tests. However, with authentic assessment, students are given a rubric in advance so that they know exactly how they will be assessed. The more complex an authentic task is, the more likely that students will have a variety of answers in a multitude of formats, making cheating unlikely.

RETAKING ASSESSMENTS

What I hear, I forget. What I see, I remember. What I do, I understand.
—Confucius

The standard protocol in mathematics classes is that teachers do not allow students to redo an assessment for a revised higher grade. Here, assessment refers to any activity for which the student will receive a grade, such as homework, quizzes, tests, and cumulative exams. There are a variety of factors regarding why teachers hold this belief and for the most part they are neither accurate nor beneficial to students.

According to Black and Wiliam (2009) formative assessment practices and feedback are two strategies that make the greatest impact on student learning. Unfortunately, assessments treated as summative are much more predominant than assessments treated as formative.

Not Like That!

Mr. Jones gives tests at the end of each unit, and the grades students earn are used to determine term grades. Despite requests to retake the test or correct problems missed because of illness, lack of preparation, or realizing a mistake right after the test is over, Mr. Jones does not allow any retakes or assessment corrections.

As a high school teacher, he is trying to prepare his students for college and the "real world." He feels that students are unmotivated to learn from any mistakes made on the assessment. He also feels that allowing retakes or changes is unfair to those students who have prepared for the test. And, last but not least, he feels that regrading will place an additional burden on his time.

Often you will hear a teacher say, "students will not be given second chances in the real world!" Of course we are given second chances in the real world. If you want a higher score on the SAT, you can retake it as many times as you want. If you teach a lesson poorly or forget to attend a meeting, you will very likely be given many additional chances to succeed. Even if you get fired from your job, you'll get another one and start over. Second chances are part of learning and part of life.

Teach Like This!

Mrs. Robins realizes that not all students will perform well at any given time on any given topic. Because mathematics builds on itself she believes that all assessments are formative up until students take a state mandated end-of-course exam or an end-of-year comprehensive test, both of which are intended to be truly summative.

Mrs. Robins has two policies regarding assessments:

1. All students are required to make test corrections for half credit. Because quizzes are meant to be formative, she analyzes the grades on each quiz to determine if there is a gap in the learning or if there was something that she didn't explain well. If students are simply being

lazy or making silly operational errors, she requires corrections for no credit. However, if she determines that the reason for the poor grades was lack of conceptual understanding, she reteaches and addresses any misunderstandings and then students make corrections for full or partial credit.

2. If a student fails an assessment (quiz or test), she can retake it for full credit or make test corrections for half credit. Before retaking the assessment, she needs to attend a tutoring session. After doing this, she is able to earn full credit for the make-up assessment.

In this perspective, students and teachers see assessment as a key component of the learning *process*, not just a product of what they know at a particular moment in time. Students, therefore are invested in analyzing the errors, mistakes, or skipped answers on their tests as they see these areas as the very opportunity to focus their learning by building that knowledge, and eventually demonstrating it again on a retake or future assessment.

Students tend to look at graded assessments superficially rather than seeing them as additional opportunities to learn from their mistakes. To most, spending time understanding their mistakes feels like a futile effort because what is done is done and they cannot get additional points for it. However, students are more likely become invested when there is a reward for their effort. Teachers tend to leave it up to the students to take advantage of additional learning opportunities, but in the new educational world of high stakes testing, it is important that teachers create and mandate opportunities for students to continue the learning process beyond an assessment.

The following is an example of a retake/corrections policy that could be given to students at the beginning of the term.

Assessment Retakes and Corrections

- If you fail a quiz or test you may attend a tutoring session and then retake the quiz or test (possibly a different version) for full credit.
- After a quiz, *all* students will be required to complete corrections. Depending on the circumstances this will not always be for credit. I will make that determination after analyzing the quiz results. *You may use your notes but not a partner when making corrections.*
- After a test all students will be required to make corrections for half credit.
- In order to receive any credit for corrections you must:

    1. Use a colored pencil.
    2. Write on the assessment itself (beside the problem or in the margins).
    3. Identify your mistake. If it is operational, circle it. If it is procedural, write out a short statement explaining why you missed it.
    4. Correct your mistake and circle your new answer.

Unacceptable corrections:

- "I missed it."
- "It is false instead of true."
- "It is *a* instead of *c*."
- "Because. . . ."
- "The calculator gave me the wrong answer."

Acceptable corrections:

- "I should have combined like terms first."
- "I confused combining like terms with the addition property."

Experience shows that when teachers use assessment corrections, students:

- Make significant gains in conceptual understanding;
- Find value in making corrections;
- Appreciate and are eager to correct their mistakes; and
- Gain trust that teachers truly care about their learning.

The idea behind assessment corrections is really very simple; however, as with most things, the devil is in the details. As teachers, we need to make sure everything is streamlined in order to maintain efficiency and optimum timing.

Most people learn best by correcting a mistake as soon as possible. Students lose interest the longer it takes to have the opportunity to make the correction. It is in everyone's best interest to grade assessments as soon as possible and give them back to students for corrections while the material is still fresh in their minds. Because we need to model for students how we want the corrections to be done, on the first couple of quizzes it is a good idea to allow them to make corrections twice with written or verbal feedback from you on how to present their corrections.

Research suggests that students who perform poorly may believe they have not received enough instruction, have not been taught in a manner they prefer, or were given material that was too difficult for them. However, we can help to combat these beliefs through assessment correction where students are required to understand where they made mistakes. They are forced to play an active role in their own learning and remedy their own deficiencies.

Assessment corrections can be beneficial to teachers as well as students. Corrections on the tests themselves can be great timesavers. They are a small investment of time that will result in great rewards in student learning.

Once students are in the habit of making corrections, they tend to ask more questions about where the mistake is made as opposed to arguing about a score because they truly want to develop understanding. This leads to a productive use of class time. Typically, a teacher hands back an assessment and opens the floor to questions or goes over each problem individually. This does not work well for unmotivated students and is generally a waste of time.

Instead, use this time as a perfect opportunity to throw in some differentiation. While some students are working on corrections you can provide challenge activities, puzzles, or ACT prep materials for students who performed well on the assessment or use the time as an opportunity for small group instruction or peer tutoring for students who lack conceptual understanding.

A policy of assessment corrections also requires that we as educators evaluate our own assessments to determine whether there is ambiguity in the questioning and whether the questions reflect what was taught on the objective. At the end of the day, the most important thing is that students are learning and that learning doesn't end with each assessment.

ANALYZING DATA

Data can play a central role in professional development that goes beyond attending an isolated workshop to create a thriving professional learning community, as described by assessment guru Dylan Wiliam. Research has shown that using data in instructional decisions can lead to improved student performance (Wayman, 2005). When it comes to improving instruction and learning, it's not the quantity of the data that counts, but how the information is used (Hamilton et al., 2009).

For any type of assessment—formative, interim, or authentic summative—it is imperative that teachers analyze the data to help guide their instruction. You can use data to express ideas in terms of numbers or graphs, but these tools have very little value unless they are used to improve instruction. Data analysis can target individual students, individual questions, or groups of questions that relate to the same standard. Simple data analysis can help educators make informed decisions by providing a quick snapshot of what students have or have not mastered.

Not Like That!

Mrs. Buckley corrects, records, and returns her students' quizzes and encourages them to review their incorrect answers. Also, if students score below a 70 percent, she tells them they should see her after school for help. She doesn't give further thought to any specific results from the assessment, but continues with her curriculum regardless of any type of assessment results. She has access to student data that will give her a graph of each student's past performance at each grade level on state standardized tests, but she feels that it is irrelevant.

Teach Like This!

Miss Roberts analyzes the data from a variety of different perspectives. As good teachers do, she identifies individual students who have struggled on the assessment, and also tries to determine the root cause of their struggles by viewing and analyzing their work. This allows her to provide specific feedback and practice material to meet the needs of individual students. She analyzes the class score on each question to determine whether she needs to reteach something to the entire class. Her data analysis also allows her to consider if a particular question is ambiguous or simply a poor question, so that she can reword the assessment in the future to avoid confusion. She also identifies common mistakes and shares all of this with her students.

Miss Roberts's department uses common assessments and she meets with her team to discuss the results of each quiz and test to determine if changes in the curriculum or the assessment could make an impact on student performance.

Data analysis is most effective when it is conducted as a collaborative effort with other teachers who have the same standards and assessments, and when meaningful dialogue can occur to address specific strategies to increase student learning in context. Identifying patterns of strengths and weaknesses allows educators to prepare for reteaching, enrichment, and intervention, and guides them in adjusting the curriculum to address those patterns for current and future classes.

Miss Roberts also looks at each individual's previous performance on state standardized tests. This allows her to see if a student is performing above, below, or at expectations. She identifies at risk students at the beginning of the term and uses that information to provide the necessary scaffolding and possible intervention. She also identifies high-achieving students who may need opportunities to be challenged in the classroom, or students who have performed well in the past but seem to be on a downward spiral and might need more of an emotional support system due to circumstances outside of the classroom.

Well-analyzed data help us to improve teaching and learning and allow us to use targeted and collaborative efforts that can have immediate and significant gains for student achievement. So if data are so important, why do teachers not utilize them effectively? Because using data has turned into a complex and broad process that includes overwhelming philosophies that are not truly applicable in the classroom. This makes teachers lose sight of the simple data analysis that can occur in a reasonable amount of time, provide relevant information, and address the immediate issues at hand. The ultimate question is not "Did the students pass the test and has our school met AYP?" The most important questions are simply "Does the student understand the objective and if not what am I going to do about it?"

Intervention has become the primary improvement strategy for trying to bring students up to par on the state objectives; however, there is only so much time in a day for intervention and only a small number of students should require intervention. Therefore, in order to facilitate progress we must turn to educators to strengthen their instruction, activities and assessments in order to meet the state standards. The only way to do this is for educators to dig in and mine the data in a way that is both feasible and understandable.

By mining the data, teachers begin to understand students more clearly and become more reflective and introspective regarding their teaching practices. Teachers do not have to become overburdened with school improvement processes that require copious, sophisticated, or special expertise. They can learn to conduct the data analyses that will have the greatest impact on learning and achievement themselves.

Over-analysis can contribute to overload—the propensity to create long, detailed, "comprehensive" improvement plans and documents that few read or remember. Because we gather so much data and because they reveal so many opportunities for improvement, we set too many goals and launch too many initiatives, overtaxing our teachers and our systems (Fullan and Stiegelbauer, 1991).

The bottom line is that improvement can be achieved through simple data-driven strategies designed to identify areas of strength and weaknesses of students through an ongoing process that changes as the needs of the students change.

KEY TAKEAWAY POINTS

- Teachers should be conducting frequent formative assessments throughout a lesson to gauge individual student and whole class understanding.
- Summative assessments, without data analysis, serve almost no purpose.
- The more opportunities to have students demonstrate what they do know, as opposed to what they do not know, through authentic assessments, the more likely you will truly know how much they understand.
- As a general practice, students should be allowed to retake assessments.
- Data driven instruction is critical to success of students. Assessing only for the purpose of having grades will not impact student learning.

Chapter Four

Relationships

No significant learning occurs without a significant relationship
—James Comer

Relationships with all stakeholders are an essential and often overlooked aspect of education. They may not be as critical as content or pedagogical knowledge, but without established relationships with students, parents, and other teachers, the job of being a successful (math) teacher is much harder. It is important to spend time building these relationships early, as they come in handy when times are difficult.

MANAGING THE CLASSROOM

Classroom management is one of the most important parts of being a successful teacher. You might have all the skills necessary to help students learn math, but just teaching math without effectively running your classroom leads to wasted effort. There are entire books and graduate courses on classroom management, but this section describes a few of the fundamental characteristics of successful management and provides some common sense tips.

Not Like That!

Mr. Moore is a passionate educator, but his classroom lacks any structure or routine. He is often unaware of the behaviors occurring during the class period as he often only focuses on one student at a time rather than observing and guiding the rest of the class. Mr. Moore often ignores disrespectful behavior, hoping that the student will stop on his own. If ignoring does not work, he blows up at the student and removes him from the class.

Parents repeatedly contact the principal to complain that students are not learning in Mr. Moore's classroom and that frequent disruptions go unnoticed. Mr. Moore gets so focused on the students who misbehave that he does not meet the needs of the silent majority. He becomes agitated and berates the offenders (and the entire class), which takes away valuable instruction time.

The sad part is that yelling is not an effective method of regaining control of the classroom, but unfortunately it becomes the precedent for the year. Students who would not typically cause behavior problems end up following the "pack mentality" and joining in on the misbehavior. This snowball effect creates a chaotic classroom where very little learning takes place.

Teach Like This!

It is important at the beginning of the year to guide students through classroom routines and expectations. Make students part of the process, so that they take ownership of the culture of the classroom. Effective classroom management looks different at various grade levels, but the similarity is that students feel safe, secure, and have a predictable routines and boundaries for classroom behavior. Here are some tips that can be tweaked to work for all grade levels.

1. Make students responsible for their own learning environment. Have them help create the classroom rules based on what they need to have a good learning environment. You can start by telling students that the rules need to cover three main categories: respectful, responsible, and resourceful behavior.

   You can guide elementary students to create the rules for each category. For middle school students, you can define each of the categories and have them break into groups and write three to five rules for each category. With high school students, have the class agree on three comprehensive rules for each category.

2. Establish rules immediately and enforce them consistently. Otherwise, you open the door for future misbehavior and negotiation of rules. It is also important that students understand the consequences for rule infractions. The younger the student, the more important it is to establish rituals and routines, in addition to rules. This helps them to feel safe and in control. Older students tend to take advantage of inconsistencies in rule enforcement. They have the ability to call teachers out on inconsistencies and use that ability to try to negotiate the rules and the consequences.

3. Use positive instead of negative language whenever possible. Instead of "Don't Talk," say, "This is learning and listening time." Give posi-

tive feedback regarding good behavior. Understand that students feed off your emotions. If you are calm, then you have a greater chance of keeping students calm. If you become angry and agitated, it is contagious and the students become agitated as well because they have lost the feeling of a safe and secure environment. This is particularly true for young students who have not learned self-control yet.

4. Actively watch for students who are gearing up for a confrontation or some type of emotional melt-down. Often, disruptive behavior can be avoided if teachers catch the small "tells" that precede these events. Avoid confrontations in front of other students. Do not argue with a student in front of the classroom. Students of all ages learn how to quickly derail instruction by attempting to argue. Go stand next to a student who seems to be heading down a disruptive path. Proximity control is one of the most effective strategies for stopping misconduct before it gets started.

5. Be prepared and organized. Whether you are an elementary, middle, or high school teacher it is important that you have well-prepared lessons, well thought out transitions and appropriate amounts of time for lessons. It is also critical to be prepared for students who learn at different paces. Bored, gifted students sometimes can be the greatest behavioral challenges in the classroom. A good idea is to have an activity station that has different types of challenges like puzzles, brain teasers, and books that they can use to occupy their time without direct instruction from you.

6. Develop relationships with parents that are based on positive aspects instead of negative. This is particularly applicable to the middle and high school teachers. It is not ideal when the first contact of the school year is caused by a negative interaction. Establishing a positive collaborative relationship first is more likely to yield support when necessary.

   A wise principal really brought this into focus for a high school when he said, "The parents have sent you the best child they have. They are not keeping the good child home for themselves." No one wants to hear only about the bad behaviors of their child. When contacting a parent, think about how you would want to be treated if it were your child. Send out positive information. Call a parent to say how proud you are of a student's progress and that she earned an "A" on her last test. Parents can be your best advocate when they feel they are in a partnership with you to better meet the needs of their child.

None of these tips are effective unless you develop a relationship with students that lets them know that you are a responsible, caring adult and that you will create a safe environment for them. Students might not remember everything you taught them, but they will certainly remember how you treated them and how you made them feel.

Relationships with students are paramount to classroom management. A caring adult relationship can change a child's life. Take a step back and analyze your current style of classroom management. If you are taking misbehavior personally, then some of the newest research on brain development and the effect of Adverse Childhood Experiences (ACEs) might be useful to you. The goal is to move from reacting and telling students what to do to asking them, "What can I do for you right now that will help you and allow me to teach?" We want to help our students learn to self-regulate. This does not mean that there are no consequences for bad behavior; it simply means that we need to help get the child out of high stress mode and into a mental and emotional place where we can be effective teachers.

John Medina has done some tremendous research on the brain that explains the effect of cortisol (called the "stress hormone") on student learning. Cortisol is released during times of stress. If a student is undergoing adverse childhood experiences and has reached toxic stress levels, the levels of cortisol become high and do not go back down to acceptable levels after the immediate crisis passes. Cortisol is what triggers the fight, flight, or freeze mechanism in our bodies and prolonged, high levels of cortisol impair cognitive performance. Children need to feel safe in order to reduce cortisol levels so they are in position to learn.

There are many schools of thought regarding classroom management. It is critical that you explore all the data and research to best meet the needs of your students. The better your classroom management becomes, the more satisfying the career of teaching will be.

BUILDING STUDENT RELATIONSHIPS

The process of building a relationship with each individual student as well as with the class is one of the most important and underrated aspects to successful teaching. As more and more research shows that adverse childhood experiences have a lasting impact on health and cognitive learning, it is more critical than ever that students have a caring adult that provides a safe environment.

Not Like That!

Mrs. True is everyone's favorite teacher. She loves to laugh, joke, and play with the students. Students easily distract her from the lesson by involving her in discussions that do not pertain to the objective, such as the latest movies and typical school drama and gossip. She has difficulty getting the class back on track because she does not want to hurt anyone's feelings and she enjoys feeling like a friend to the students. She also struggles to make students follow the rules because they do not respect her as an authority figure. As the year progresses she has more and more problems with students arriving late, not completing assignments, and asking for help during quizzes and tests because they have not studied.

At the beginning of the following year when standardized test scores are released, two of her students discuss the decline in their scores compared to previous years. In a moment of honesty and clarity, one student says, "I love Mrs. True and I enjoyed her class all year, but I am actually dumber for having taken her class." This student, a future National Merit Scholar, feels, after the fact, that Mrs. True did him a disservice by trying to be a friend to the students rather than a teacher who ensured they reach their academic potential.

On the opposite end of that spectrum, Mr. Thigpen is a very strict teacher who does not develop a personal relationship with his students. Students are rarely off task and the classroom is extremely quiet. There is little interaction between students and only academic interaction between student and teacher. If a student comes to class without supplies, he humiliates this student in front of the class. Students do not feel comfortable asking questions or seeking help when they don't understand. This is an unpleasant classroom environment that inhibits student participation and learning.

Teach Like This!

Mr. Williams, a tenth-grade math teacher, stands at the door each morning and greets students by name as they enter. He asks them about their lives and different extracurricular events they attend. At the beginning of the year he has each student fill out an information sheet, labeled as a time capsule, which gathers information about their interests, learning style, feelings about math, and more. He takes a picture of the class and saves it along with the time capsules.

This simple activity gives him a tremendous amount of information about the students that allows him to build an appropriate relationship based on personal and academic knowledge of the student. On the day of their high school graduation, he leaves their "time capsule" in their seats along with a copy of the class picture and a sweet note wishing them well for the future.

Mr. Williams sponsors the math team and attends lots of different sports games and other school events to show his support of the students and his school. During class time, Mr. Williams spends two to three minutes discussing current events and commenting on student participation in various events or honors that they have received, and then he begins class. Once he begins class he does not allow students to steer the direction of the class back into irrelevant topics after the initial "warm up."

If a student does not complete assignments or grades begin to drop, Mr. Williams refers back to the information sheet and has a conference with the student to discuss what is causing the lack of motivation. The relationship that he has spent time building allows the student to feel comfortable sharing thoughts and coming up with a plan of action to get back on track. This does not negate any consequences that the student must face, but Mr. Williams makes sure that the student feels supported in a safe environment.

Students do not need to view the teacher as a friend, but they do need to see the teacher as a mentor, role model, and most importantly, an advocate who cares about them as a whole person, not just academically. When students feel championed by their teacher, they will work harder and they will be motivated to perform well for you even when they don't want it for themselves.

For example, Ms. Roberts teaches a very low-level math class that includes inclusion students. No one expects them to perform well on standardized tests. However, year after year her students outperform expectations. She rarely experiences many behavior issues either. When someone asks how she does it, she always responds, "The first thing I teach them is that I care about them and the second thing I show them is that they can be successful at math."

The plan is simple: Ms. Roberts starts off with easy assessments and provides help and hints so that each student achieves a high score. She spends individual time, even just a minute or two, with each student every day helping with the math. As the year goes on and students become more confident in their abilities, she scales back the scaffolding until they rarely need help and can complete assignments and assessments by themselves.

It is important to develop an individual relationship with each student to help them feel comfortable and safe. A good solid relationship allows the teacher to draw upon that relationship in case there is an issue with discipline or their work. Students typically spend more time with teachers than they do any other adult. It is critical that we do not take that responsibility lightly. We may be the only good adult role model currently in a student's life.

According to Maslow's hierarchy of needs, students cannot engage in learning until their physiological needs (food, water, warmth, and rest) and safety needs have been met. Many schools have programs that provide students with clothes, food, snacks, and supplies. However, the older students get, the more embarrassing it becomes to let people know that they need something. This is where having a good relationship will allow you to (1) know that a student has needs, and (2) speak to the student privately and arrange a non-intrusive way to get them the support they need. We need to preserve the dignity of our students.

Students' long-term mental health is shaped by their experiences as children. "Thanks to the plasticity of the developing brain and other biological systems, the neurobiological response to chronic stress can be buffered and even reversed, especially when we intervene early in children's lives" (Thompson, 2014, p. 41).

Dr. Ross Thompson, distinguished professor of psychology at the University of California, writes: "In particular, warm and nurturing relationships between children and adults can serve as a powerful bulwark against the neurobiological changes that accompany stress, and interventions that help build such relationships have shown particular promise" (Thompson, 2014, p. 41). An increasing amount of research shows that adverse childhood experiences affect cognitive learning and lifelong health; teachers can alleviate some of that stress by providing a place of safety and comfort. Teachers can offer a place where students believe that they are cared for and that adults have their best interests at heart. When teachers can do this, students learn.

REACHING OUT TO PARENTS

It is clear that parental involvement can make a difference in a child's educational success. Parents can either be your best supporter or they can be your biggest frustration and a hindrance to student achievement. How you communicate and bridge the gap from school to parent can determine the success of your relationship with both the student and the parent. The challenge lies in how to create the right kind of involvement and which activities are truly worthwhile. As educators, we want to be proactive by anticipating needs that can be met by parental involvement in helping a child succeed, as opposed to being reactive and informing parents after the fact.

Parents from all economic and cultural backgrounds want their children to succeed, and most want to be active in their child's learning even if they cannot attend all the fundraising and PTA/PTO meetings. Parents need our input as to how they can best participate in their child's learning regardless of limited time, education, or resources on their part. Our job is to create a partnership with the parent that will have the most impact on the student's achievement.

Not Like That!

At the end of term one, after all the grades have been submitted, Mr. Love sends out a general email letting parents know their child's end-of-quarter grade. Included is a list of assignments and the student's scores for each. Mr. Love has not been in contact with any parents before this moment except to deal with discipline issues. He does not feel that it is necessary to keep the online grading program updated because he hands back graded work to students and it is their responsibility to keep up with it.

When a parent contacts him wanting to know why their child is failing, he provides a list of missing classwork and homework and informs the parent of failing grades on quizzes and tests. Unfortunately, Mr. Love has reported all the facts but has not offered any strategies or solutions and the quarter 1 grade is now a part of the student's permanent record. In this scenario, Mr. Love is taking a reactive stance in which he has told the family about events that have already occurred without allowing for collaboration or growth.

The parents, who are naturally upset, contact the administration to request a meeting. They want to know why they were not contacted before the failing grade was submitted so that they could have made an attempt to either motivate the student or explore options for tutoring or remediation.

Teach Like This!

Mrs. Dalton provides an information sheet to each parent and student at the beginning of the school year that includes her contact information, directions for how to access the online gradebook program and information regarding tutoring and remediation services that are offered through the school. In turn, she asks the parents for their contact information. She takes the time to set up an e-mail group and a one-way text messaging system that allows her to contact all the parents at once. When she updates the online grading program, she contacts all the parents letting them know that there are current grades online.

Making contact with parents allows her to address a number of scenarios throughout the year. For instance, Mandy is doing well on daily work, but fails the first test. Mrs. Dalton e-mails Mandy's parent to set up a phone conference, which she starts by praising how well Mandy has been doing on the daily work. She goes on to express her concern over the failed test. In talking with the parent, she discovers that Mandy suffers from test anxiety, as she suspected. Mrs. Dalton asks the parent what strategies have helped Mandy in the past.

In the end, they decide that Mandy will come to Mrs. Dalton before school on the day of a test to ask questions about the review, work any problems that she didn't understand, and build her confidence. Mrs. Dalton also agrees to stop by Mandy's desk during the test, not to tell her how to work a problem, but to reassure her on a few problems that she is working right. They also decide that if Mandy is having a particularly stressful time with the material, she can stay after school and take the test without other students around.

Another student, Jordan, starts off the nine weeks doing extremely well. However, as the term progresses, his grades steadily sink. Mrs. Dalton contacts his parent and asks for a phone conference. She expresses her concern about the decline of Jordan's grades and asks the parent what she thinks the issue might be—not understanding the material, lack of effort, personal issues?

His mother shares that the family is struggling with a possible divorce and it has really affected Jordan. At the end of the conversation, they have a plan in place for Jordan to attend an after school tutoring program to catch up on key concepts. He can then retake some assignments and assessments. Mrs. Dalton takes the time to speak with Jordan one-on-one and expresses sympathy for his situation. She also introduces him to a guidance counselor who is able to gain his trust and gives him a place where he can talk out some of his feelings and frustrations.

Jeremiah is a typical high energy young man who is the life of the classroom. However, he never stays focused on the task at hand, and although he is not failing, his grades are not where they should be. Mrs. Dalton looks up his math grades from previous years and notices that he is not performing up to his potential in her class. She also speaks to some of his other teachers and asks if they are noticing the same thing. She e-mails Jeremiah's mother and asks to meet with her during her plan time.

When they meet, Mrs. Dalton says that she is concerned that Jeremiah is not doing as well as he should. His mom explains that he has ADHD but no longer wants to take his medication, so she allowed him to begin the school year without them on the condition that his grades did not drop. Together, they bring Jeremiah into the conference and explain that in the interest of his grades, he should go back on his medication.

In each scenario, Mrs. Dalton makes sure to praise the child and show concern for his or her well-being. She has a plan ready to propose to the parents and yet the parents still feel like they are a part of the decision-making process. When contacting parents, a good rule of thumb is to treat them like you would want to be treated. No one likes to hear only the bad things about their child.

Not every child learns the same way, and not every parent communicates in the same format. As a teacher, you need to be well-versed in many forms of outreach, such as e-mails, Web site information, phone calls, social media, and newsletters. Also, personalized communication should not occur only when negative issues arise; you can also share positive achievements such as a high quiz score or on-time completion of the last three homework assignments.

It is important to be proactive and involve families early on, should you notice that a student is struggling with classwork or participation. Not only will families appreciate the dedication, but many times they will offer insightful suggestions on how to help motivate the student. In addition, this helps build a positive relationship so that if the students do continue to struggle with performance, the parents sees you as an ally and resource, rather than as the task master who is "giving" their child a poor grade.

At the elementary level, parents invest a lot of time helping with homework and volunteering at the school. Therefore, teachers often get to know the parents and forge strong communication networks. However, at the high school level parents often feel less equipped to help with homework and there are not as many opportunities to volunteer. The sad part is that because of a lack of communication, most parents do not know how to help a student or what resources are available to their student through the school.

Effective parent involvement starts with information. Schools can offer workshops on attendance, homework help, and general parenting skills to provide a support system for children of broken and blended families. They can also use contracts between the student and the parent which helps to commit parents to actively participating in their child's learning. For instance, communicating with families regarding attendance and asking them to sign a contract committing to getting their child to school on time can reduce chronic absenteeism at the elementary, middle, and high school levels.

This begins a true partnership between school and parent. However, even if this school-level infrastructure is not an option, as individual teachers we can inform parents as much as possible regarding all resources that will benefit their child. In the current educational climate, where graduation rate is so important, some schools have instituted policies under which a teacher cannot fail a student unless they have documented parental contact, intervention attempts, and a guidance (graduation coach) referral.

Using parental involvement to solve a particular problem can be very beneficial; however, ultimately it is a support and not a substitute for good teaching practices. Just because a student does not have a parental support system, we cannot and must not give up on that child. You may be the only role model and cheerleader that student has ever had. Make it count!

FOSTERING PROFESSIONAL LEARNING COMMUNITIES

Professional Learning Communities (PLCs) is one of the most frequently misunderstood buzzwords in education. A PLC, as defined by author Richard DuFour, is an ongoing process in which educators work collaboratively in recurring cycles of collective inquiry and action research to achieve better results for the students they serve. Professional learning communities operate under the assumption that the key to improved learning for students is continuous job-embedded learning for educators (DuFour, 2004).

In simpler terms, a PLC fosters collaborative learning between grade level and content area teachers and other professionals such as Response to Intervention (RTI) coordinators, learning leaders, counselors, special education teachers, and administrators. Districts and schools use the PLC frame-work to organize teachers into working groups.

Not Like That!

The school district has mandated that all schools begin a PLC where each school in the county will work with their cohorts to create a scope and sequence based on the current math standards. After they create the scope and sequence, they will develop common assessments.

In the first PLC, everyone is designated a role—time keeper, recorder, and such. Everyone finds this to be a senseless practice and treats it like a joke. They have a problem—the three different schools use three different curricula and textbooks. No one can agree on a sequence because each already has a familiar (and therefore perfect) scope and sequence. They also lack a common planning time and can't correct this because of student scheduling.

In order to satisfy the administration and have something to put on the district Web site, they find a scope and sequence from another school, tweak it a little, and submit it with no intention of ever actually implementing it. They spend the rest of the PLC time having a gripe session about the standards, testing, and teacher evaluation process and feeling frustrated that this common scope and sequence is one more thing that is being asked of them that can't feasibly be implemented.

Teach Like This!

The PLC approach is a comprehensive program that can take years to fully implement within a school. What starts off as a great concept that would certainly be in the best interest of students, departments, schools, and districts soon gets dropped because the task seems enormous. In this case, the top-down approach can be overwhelming and unworkable, especially without faculty buy-in, but working from the bottom up lets small, easy to implement changes create momentum.

The district's PLC meetings are ineffectual and discouraging. But Mr. Singh starts collaborating with another Algebra teacher, Ms. Wheeler, who is admired for her approach to teaching and the successful relationships that she has with her students. They have very similar styles of teaching and are both dedicated to continuing their own learning process.

They start with goals for the first year that they know can feasibly be implemented:

1. They sequence their units and set a timeline for completion of those units.

2. They work one unit at a time and put their resources together to use common materials, notes, and activities.
3. They create common assessments.
4. They meet in the hallway during break, lunch, or after school for short periods of time to discuss teaching strategies and reflect on where they need to edit, add, or delete materials based on student understanding and performance.

Because they do this in small increments, it takes minimal time. It is amazing what ten to fifteen minutes of PLC time used on a regular basis can accomplish. The teachers see great results in student achievement in their very first year just by implementing those steps.

The second year, they bring the entire Algebra 1 department of five teachers on board. They meet for two days in the summer to share results and strategies and tweak their pacing and curriculum, with each educator bringing to the table a resource or strategy that has worked to increase student understanding. Everyone is treated as a valuable member of the team with something to contribute. Each teacher has one or more "superpowers" such as leadership, organizational skills, data mining, Internet resource mining, classroom management, special needs experience, or technological know-how.

When they put all that talent together, they become one of the top math departments in the state with very high achievement scores in Algebra 1. The rest of the math department joins in; each subject area bands together and starts doing what Mr. Singh and Ms. Wheeler did during the first year. They also start mining individual student data, looking at past classroom grades, standardized test scores, behavioral issues, and motivational techniques that have worked. The team regularly connects to special education teachers and guidance counselors. Fortunately, their administrator believes that it is his job to provide this successful project with the necessary tools, offer a support system, and stay out of the way of the freight train as it gathers momentum.

By the third year they meet as a department and discuss how to seamlessly transition students from subject to subject using common vocabulary and strategies. For instance, Algebra 1 teachers agree to use geometric figures and calculate area and perimeter when they teach combining like terms and distributive property. Algebra 2 and Algebra 1 teacher agree on methods of factoring using common vocabulary and strategies so that students will make the connections to prior knowledge more quickly. As an additional goal, they also agree on standards that will enhance SAT scores even though these are not a priority in the state standardized testing.

During this year, they receive a state mandate to provide RTI at the high school level, which requires them to administer testing; place students in Tier 1, 2, or 3 categories; and provide intervention with data points. What results is an intervention period in which students receive an appropriately leveled remediation such as basic skills remediation, standards remediation, or enrichment like SAT prep. Because a student may have one math teacher for class and another for intervention it is very important that they work together to further the learning needs of the child.

The two main results that the teachers hope to see at the end of this year are that struggling students, particularly those in Tier 3, will have stronger basic skills such as operations with integers without the use of the calculator, and those academically stronger students will have higher scores on the ACT or SAT.

Professional learning communities require a long-term commitment so that the practices become routine. An appropriate amount of time must be allotted for good practices to develop. Ultimately the focus is on students, and on generating products and strategies that improve student achievement at the group, subgroup, and individual levels. Success depends on the professionalism of teachers, who must be prepared to work collaboratively and take advantage of each other's strengths. PLCs need to have a clear goal that can be attained by a series of steps, as follows:

1. Identify an issue that is hindering student achievement.
2. Select evidenced-based strategies that are implementable in your school culture.
3. Develop a common scope, sequence, and assessment that will demonstrate learning.
4. Analyze the results as they pertain to groups, subgroups, and individuals.
5. Adjust pacing, curriculum, assessments, and student relationships based on the evidence of your analysis.

Wash, rinse, and repeat for each goal as determined by your classroom, school, district, and state, and pretty soon you will have attained amazing results. You will have the added blessing of knowing that you are not in this alone and you have others with whom to share successes and failures.

Professional development at the school or district level is typically so broad that teachers leave with more questions than answers, or they are given mandates without a plan for reaching those goals. This is why a well-developed PLC is a critical component for success. This PLC process fosters the birth of natural learning leaders within each school and department, which provides an amazing resource for the school and district. It also allows teachers who have developed effective learning strategies to share them with new teachers or struggling teachers in a manner that is designed to build up and not tear down.

Taking on every problem in education at once is an insurmountable task, but we can take it one bite at a time and begin at the bottom of the ladder with the things that are within our control. Bring a few collaborators on board and the next thing you know you are at the top of the ladder in terms of student achievement and you have created your own PLC. In the process, you will have gained an education from your fellow richly experienced and talented educators that college courses could never have provided.

COLLABORATING ON SOCIAL MEDIA

To grow as educators after we earn our university degrees, we read a few books and obtain professional development from two primary resources— conferences and school in-service. Conferences are a wonderful way to net-work and to meet other educators but can be so costly and time-consuming that they are not always feasible. School in-service professional development has the benefit of being provided by the district, but is typically too generic to meet the needs of all the professional educators teaching in a wide variety of content areas. An advantage of conferences and in-service is the ability to collaborate face-to-face with other educators. However, a concerning draw-back is that they provide a very limited number of people with whom to collaborate.

If we want our students to be life-long learners it is important that we model for them how that can be achieved. In the technological age, this is easier than ever before. Professional discussions and collaborations are no longer exclusively occurring in person. Social media and the Internet provide opportunities for collaboration and growth as an educator. The amazing thing about the twenty-first century is that we have so much knowledge available instantaneously at our fingertips through technology. We have access to edu-cators all over the world!

Not Like That!

Mr. Thomas has taught calculus for the last twenty-two years. He has his resources, materials, and a mindset that "if it is not broken then why fix it." By his standards, his students are successful. He does not participate in any collaboration with the rest of the math department and refuses to change his material or methods to "teach to the test." His students need outside tutoring in ACT/SAT preparation and technology skills. While his focus on content is admirable, juniors and seniors are taxed with the burden of trying to get into college and find a way to pay for it. ACT/SAT scores can be the key that opens those doors. We do students a disservice by not incorporating test-taking skills along with content.

The sad part is that this brilliant man has so much to share with students and teachers alike if he would participate in collaborative discussions. Through collaboration, he could learn to better explain connections across standards and why certain strategies are better than others for teaching certain content.

For instance, eighth-grade math teachers and Algebra 1 teachers typically teach students to put a linear equation in slope-intercept form before graphing. Students get so stuck in that mode that they cannot derive information directly from the standard form, which actually gives more information with less work. It also sets the baseline for working with other polynomials in standard form. When you get set in your own world, you only teach what students need to know for you, without considering the next step in their educational journey.

While Mr. Thomas is an expert in content material, he could benefit from other teachers' knowledge regarding current resources, technological advances, and higher education needs that go beyond content.

Teach Like This!

Looking at the data at the end of the school year, the math department recognizes that although the end-of-course math scores are good and there is growth in student achievement, ACT scores are lower than they should be. In the current school culture, the ACT is not big deal; however, new guidelines will use ACT achievement scores for school and teacher evaluation purposes. ACT scores need to become a greater focus.

At the end of the school year, the department meets and decides to spend the summer educating themselves on everything ACT, including content material, best test-taking strategies, changes in format, and strategies to motivate students. Two teachers go to a conference on ACT strategies and two teachers purchase several of the best-rated ACT books and summarize the best test-taking strategies for each content area (not just math) and create handouts for teachers throughout the school. Another teacher uses Feedly (an app that helps to keep track of blogs) to read hundreds of blogs related to the ACT and creates a list of resources for students and teachers. Everyone agrees to mine the Internet for information and real-time collaboration groups and come back with the best media resources, such as who to follow on Twitter, Facebook chat groups, and blogs that will continue to provide current information.

Through a variety of online collaborations, the team is able to assemble ten years of retired ACT tests. Using feedback from multiple online sources such as articles, blogs, Twitter, and Facebook the team finds several online resources that enable students to practice the ACT for free and provide educators, particularly teachers who do not teach a core class, with materials to use for ACT prep. The more technologically savvy teachers search hashtags and follow links to "experts" who are willing to share dropboxes full of years of materials, research, and tips. The technology maze eventually leads to a state educator who raised his school's average ACT score to a very high level, and he agrees to meet with some of the teachers at this school for a brainstorming session.

The math department meets as a PLC with an administrator and representatives from the science and English departments right before school starts and everyone pools information and resources. They create a game plan for how to implement the goal of raising ACT scores in which every teacher in the school can participate without overlapping content and strategies.

This is just one example of how technology can be used not only to advance your work as an educator but to positively impact your entire school and culture.

Students are media creatures. They learn and think through social media and for a change we can let them teach us. We can self-educate at an exponential rate when we utilize social media and the Internet for collaboration; and best of all it is usually free. It doesn't matter what your objective is— understanding a standard, enhancing classroom management strategies, motivating students, collecting materials, or raising student achievement; there is a wealth of information out there provided by experts who are willing to share their knowledge with you via social media. You can follow and discuss topics with different educators on Twitter; belong to Facebook groups that are dedicated to sharing curriculum resources; search for collaborative groups that post under certain hashtags like #msmathchat, #mtbos, #elemmathchat, or #growthmindset; and participate in live chats or watch webinars. The world of education becomes endless when we step outside of our bubble and into the world of social media.

KEY TAKEAWAY POINTS

- It is essential for a teacher to have effective classroom management skills in order to be a successful teacher. Organization and consistency are two of the best ways to develop these skills.
- Building a relationship with each of your students will yield a huge impact with their attitude, effort, and performance.
- Reach out to parents early and often and find ways to communicate about the great things that their child is doing and not just the problems.
- True and authentic Professional Learning Communities can provide some of the best professional development for teachers.
- Do not hide from social media. Embrace it as a way to communicate with other math teachers, parents, and the community.

Conclusion

Call to Action

When teachers stop learning, so do students.

—Jim Knight

We can all still learn and grow in the way we prepare and execute our lessons. Technology is always changing and new ways to engage students and help them learn mathematics are always occurring. The four major ideas outlined in this book: planning, pedagogy, assessment, and relationships are all critical to the success of a teacher, especially in mathematics. In addition, they are all connected and can lead to improved student learning.

Select (at least) one topic discussed in this book of interest to you and act upon it. Begin a cycle of inquiry to learn more and make changes to your mathematics classroom. Do not start this next week, next month, or next school year; begin the process *now*.

1. Start by reflecting on what you want to learn more about. Examples might be "How can I use my assessments to drive my instruction?" or "How can I find time to devote five minutes to summarizing my lessons at the conclusion of class?"
2. Reread the section of this book that discusses the idea that you want to explore and then reach out to other resources. Colleagues, articles, books, and the Internet are all great places to begin to learn more on this one area of focus.
3. Develop a plan of how and when you will implement the changes. If it is having better communication with parents, it is not possible to contact every parent via phone within a week, but rather you could make

two phone calls a day to reach out to families to just share some good information about their child and the course that you are instructing.

4. After a month or two, reflect on what you have done and determine the impact that it has on your students. Answer your reflective question and determine what impact it has had. Ultimately everything comes back to student learning and the change that you have made should impact it in some manner.

5. Either extend the plan further in one particular area or now attack a new topic from the book. If you focused on communication with parents via phone and it has worked well, the next step could be to set up a Web page or social media account to extend the communication. Or you may want to turn your attention to the homework that you assign. Does each forty-minute homework assignment provide benefit to your students? If not, how can you adjust, cut back, or eliminate?

Additional resources that builds off the ideas of this book include:

Pi of Life: The Hidden Happiness of Mathematics, by Sunil Singh
Mathematical Mindsets, by Jo Boaler
The Power of Mathematical Visualization, by James Tanton

References

Banilower, E. R., Smith, P. S., Weiss, I. R., Malzahn, K. A., Campbell, K. M., & Weis, A. M. (2013). Report of the 2012 national survey of science and mathematics education. Chapel Hill, NC: Horizon Research, Inc.

Boaler, J., & Dweck, C. S. (2016). *Mathematical mindsets: Unleashing students' potential through creative math, inspiring messages and innovative teaching.*

Black, P. J., & Wiliam, D. (2009). Developing the theory of formative assessment. *Educational Assessment, Evaluation and Accountability, 21*(1), 5-31.

DuFour, R.P. (2004) What is a professional learning community? *Schools as Learning Communities,* 61(8), 6-11.

Dweck, C.S. (2006). Mindset: the new psychology of success. New York: Random House.

Fullan, M., & Stiegelbauer, S. (1991). *The new meaning of educational change.* 2nd ed. New York: Teachers College Press.

Hamilton, L., Halverson, R., Jackson, S., Mandinach, E., Supovitz, J. A., & Wayman, J. C. (2009). Using student achievement data to support instructional decision making (No. NCEE 2009-4067).Washington, DC: National Center for Educational Evaluation and Regional Assistance, Institute of Education Sciences, U.S. Department of Education.

Panasuk, R. M., Stone, W. E., & Todd, J. W. (2002). Lesson planning strategy for effective mathematics teaching. Education 122, 4, 808-826.

Thompson, R.A. (2014). Stress and Child Development. *The Future of Children,* 24(1), 41-59.

Wayman, J. C. (2005). Involving teachers in data-driven decision-making: Using computer data systems to support teacher inquiry and reflection. *Journal of Education for Students Placed at Risk, 10*(3), 295–308.

Wiggins, G.P. (1993). *Assessing student performance.* San Francisco, Ca: Jossey Bass Publishers.

Wiggins, G. P., McTighe, J., Kiernan, L. J., Frost, F., & Association for Supervision and Curriculum Development. (1998). *Understanding by design.* Alexandria, Va: Association for Supervision and Curriculum Development.

About the Author

Dr. Matthew Beyranevand is a K–12 Mathematics and Science Department coordinator, an ambassador for the Global Math Project, supporter for the With Math I Can campaign, and a member of the Massachusetts STEM Advisory Council. He also serves as an adjunct professor of mathematics and education at the University of Massachusetts at Lowell and Fitchburg State University. Through his Web site, www.mathwithmatthew.com, he provides visitors with his podcast, blog, math music videos, a pilot TV show, and more resources to help increase students' curiosity and engagement in learning mathematics, while also building conceptual understanding.

Connect with Matthew

Email: mathwmatthew@gmail.com

Website: www.mathwithmatthew.com

Twitter: @mathwithmatthew

Facebook: www.facebook.com/mathwithmatthew

Instagram: math_with_matthew